STEAMERS OF THE THAMES AND MEDWAY

P.S. "Woolwich Belle"

Frontispiece

STEAMERS OF THE THAMES AND MEDWAY

BY

FRANK BURTT

AUTHOR OF

"CROSS CHANNEL AND COASTAL PADDLE STEAMERS"

ROADMASTER PUBLISHING

BAGGINS BOOK BAZAAR

First published in 1949 in Great Britain by
Richard Tilling
106 Great Dover Street
London, SE1

This edition reprinted in 1997 by
Roadmaster Publishing
PO Box 176
Chatham
Kent, ME5 9AQ

in association with
Baggins Book Bazaar
19, High Street
Rochester
Kent, ME1 1QB

ISBN 1 871814 46 4

British Library Cataloguing in Publication Data

A catalogue record of this book is available from the British Library

Printed and bound in Great Britain by
Bookcraft (Bath) Ltd, Midsomer Norton, Avon

FOREWORD

In 1934, I wrote a volume on Cross-Channel and Coastal Paddle Steamers. This present publication is a companion book dealing with the development of Passenger Steam Boats confined to the Thames and Medway and the Estuary. Over ninety per cent. of such craft is confined to the paddle means of locomotion which has hitherto been found preferable for river work although in recent years some very fine screw vessels have been placed in service. As paddle steamers have a greater fascination for the eye, and a very warm place in my affections, the final chapter is devoted to the paddle tugs of a famous firm of tug owners. In conclusion I would like to place on record the assistance given me by Lloyds' Register of Shipping, Messrs. W. G. Tilling and H. A. Allen and Major E. W. Taylerson, M.I.Mech.E.

F. BURTT.

Caradoc,
South Way,
Lewes.

CONTENTS

LIST OF ILLUSTRATIONS

CHAPTER I.

EARLY STEAMBOATS.

OLD FATHER THAMES is the most important waterway of the British Isles and a really comprehensive account of the passenger steam vessels that have operated on its course has never been published. The author has endeavoured, therefore, as far as possible to remedy that omission. A good deal of research has been necessary and, as so frequently happens, early records are either non-existent, or when found do not always agree. The author is in no position to dogmatise but has accepted as the principal authorities—The History of the Waterman's Company; Messrs. Sandham, Carson, and Ravenhill who contributed papers on Early Thames Steamboats to the Institutions of Mechanical and Civil Engineers and Naval Architects; from these publications he has quoted freely. Acknowledgment is also made to G. Aylmer for quoting from his contribution to *The Engineer* dated 9th September, 1932.

The author can trace no authentic record of any steam craft on the Thames prior to 1814. In 1813, a Mr. Dawson is supposed to have run a small steamboat between Gravesend and London. Nothing definite can be traced, and recognized authorities do not give the legend much credence, but in 1814, a Mr. Laurence did run a steamboat between Bristol and Bath, which was transferred by inland navigation to the Thames. The opposition met with from the Thames watermen obliged the little vessel to return to Bristol.

The early steamboats were not at first a success. Interruptions continually took place. The old watermen did everything they could to obstruct. Machinery frequently broke down. The working expenses were heavy. But the end was inevitable and in 1834 the last of the tilt boats,* the DUKE OF YORK, was withdrawn.

MARJORY 1814/1815.

The MARJORY is generally accepted as the first passenger steamer to work regularly on the Thames; she was built of wood in 1814, at Dumbarton by William Denny,† founder of the famous firm of W. Denny and Brothers, Dumbarton.

The MARJORY was built to the order of Mr. Anderson, a Glasgow merchant. She was of 70 tons burthen, 63 feet in length and had an engine of 14 n.h.p.

After running on the Clyde she came via the Forth and Clyde canal to the Thames in January, 1815, and ran between London and Gravesend making the passage between the two places on alternate days; she is said to have continued to ply for a few months, during which time, it was no uncommon thing for her to lay up for several days at a time for repairs. At the end of her first season she was withdrawn.¶

---

* The special name given to the large rowing boats that plied between London and Gravesend fitted with a "tilt" or awning.

† These particulars are accepted as being correct and not as shown in the list. Pages 18 and 19.

¶ Aylmer quotes her as running up to 1858.

The MARJORY took up her running on 23rd January, 1815, according to the following notice:

> "The public are respectfully informed that the new London steam engine packet MARJORY, Captain Cortis, will start precisely at ten o'clock, on Monday morning, the twenty-third inst. from Wapping Old Stairs, near the London Docks, to Milton, below Gravesend, and will return from thence at the same hour on the succeeding morning to the same stairs, the said packet having superb accommodation. Passengers and their luggage will be conveyed to and fro with more certain speed and safety, than by any other conveyance by land or water, and on reasonable fares. Passengers are requested to be punctual to the time specified. Fares, four shillings in the chief cabin, and two shillings in the fore-cabin."

At Gravesend passengers landed and embarked in rowing boats. The Waterman's History states that "Proceedings under the Watermen's Act were taken against the Captain for navigating the vessel, and his place was then taken by a freeman of the Waterman's Company."*

THAMES (1815) *ex* ARGYLE.

In the same year—1815—another steamboat named the ARGYLE came to the Thames; she is reputed to have been built in 1813, by J. & C. Wood.

This little craft came round from the Clyde via Land's End, putting in at Dublin, Wexford and Portsmouth,

* "The History of The Waterman's Company. Vol. III, 1874."

after rounding the North Foreland she is recorded as doing the 90 miles from Margate to Limehouse—with the tide—in nine hours.

On coming to London this vessel was re-named THAMES. She was 65 feet in length having a beam of 14.5 feet, 74 tons burthen, fitted with a 16 n.h.p. side-lever engine constructed by James Cook of Glasgow. The paddle wheels were 9 feet diameter, 4 feet wide, with 16 floats in each wheel, 15 inches broad. Her average speed was 6 knots.

On her journey to the Thames she encountered very bad weather. Twice during the trip boats put off from the shore, under the impression the vessel was on fire.

The THAMES made her first run to Margate on 3rd July, 1815, but in consequence of the introduction in the following year of a larger vessel—the REGENT—the THAMES was withdrawn and placed on the London-Gravesend service. The Waterman's History says at this time the MARJORY was withdrawn, re-fitted and went to France, to be employed on the Seine.*

The following somewhat amusing advertisement was issued by the promoters:

GRAVESEND STEAM PACKET
THE THAMES
CAPTAIN PAYNE,
will leave
The Tower Stairs for Gravesend.

* In 1816, a steam-boat named MARGERY was built on the Thames for service on the Seine and re-named ELISE. The author is of opinion these two vessels should not be confused.

Every morning at 8 o'clock (Friday excepted) and return every afternoon at 3 o'clock the same day, for London. Fares on Sundays, 3s. each; other days, Best Cabin 3s., Fore-cabin 2s. Persons under 14 years of age half price, and children under 3, free.

Ladies accompanied by Gentlemen in large parties of pleasure, passage free, Sundays only excepted. Refreshments provided: Tea, Bottled Porter, etc., etc. Passengers bringing refreshments will be assisted with Boiling water and Milk, provided in readiness aboard.

The Times of 8th July, 1815, refers to the THAMES in the following words:

> " She is rapid, spacious, and indeed a splendid vessel. Her cabins are large and are fitted up with all that elegance could suggest or that personal comfort could require. She presents a choice library, and back gammon, draught boards, and other amusements are provided. For the express purpose of combining delicacy with comfort, a stewardess tends on the fair sex."

The THAMES carried a very tall smoke-stack which did duty as a mast on which a large square sail was set when the wind was favourable. A long outside gallery on which the cabins opened gave the impression that she was larger than she actually was, and with her row of gun ports she presented a formidable appearance.

RICHMOND. 1814.

The RICHMOND was built at Yarmouth by Lepingwell. According to the Waterman's History the date is given as 1814, Sandham gives the year as 1818. As her boiler burst in 1817, no doubt 1814 is correct.*

She was a vessel of 50 tons burthen fitted with a 10 n.h.p. engine by Maudslay. She worked between London and Richmond. On 28th June, 1817, her boiler burst—it had only recently been repaired—just above Westminster Bridge while returning to Richmond, the "conductor" of the boat was very badly injured as well as two other persons in charge.

Mr. George Dodd in his work on the steam engine and steam packets, 1818, records that two of his Thames steamers, the LONDON and the RICHMOND carried on their station within four months 10,000 passengers. He describes his plan for lowering the chimney of these steamers to enable them to pass under the bridges; and records his proposal of water-tight iron bulkheads for dividing wooden vessels into compartments.

DEFIANCE. 1815.

During 1815 a packet named DEFIANCE was put in service between London and Margate, but no further information is available.

HOPE. 1815.

The HOPE built by Laurence at Bristol in 1815 and engined by Johnson, was put in service between London, Sheerness and Chatham. She was a vessel of 45 tons burthen and had a 6 n.h.p. engine.

* The year 1814 is confirmed in Capt. N. W. Kennedy's "Records of the Early British Steamships."

REGENT. 1816.

In 1816 the REGENT was put on the Margate service and as previously stated replaced the THAMES. She was the first steamboat to be built on the Thames. Designed by Isambard Brunel, built by Courthope, of Rotherhithe, and engined by Maudslay, the founder of the famous firm of marine engineers, Maudslay, Son and Field.

She was a beamy little vessel of 112 tons burthen and her engines of 24 n.h.p. were independent, each driving its own paddle, with the boiler placed between them. Her great fault was that, in the desire to obtain the high speed which was demanded by her passengers, everything was built lightly. It is surprising that she finished her career by being burned instead of falling to pieces.

It was on 2nd July, 1817, whilst off Whitstable that the REGENT was burnt out, having on board about 40 passengers and a crew of 10, all of whom were saved by boats putting out from the shore.

CALEDONIA. 1816.*

The CALEDONIA, the next steam vessel to run on the Thames, was built by J. & C. Wood, Glasgow, in 1816 and brought round to the Thames to operate on the Margate service. The CALEDONIA was 94 feet long b. p., beam 15.25 feet, depth of hold 9 feet, fitted with two engines built by Boulton & Watt, one engine for each wheel. They were together of 32 n.h.p., the cylinders being 23 inches diameter with a piston stroke of 30 inches. The paddle wheels were 13 feet diameter. Tons 102.

* Muirhead in his "Life of Watt" says she was launched in 1815.

James Watt, the son of the inventor of the condensing steam engine, purchased her in April, 1817, and after refitting her, left Margate on October 15th for the Scheldt. From Rotterdam on 23rd October he took her up the Rhine, arriving at Coblenz on 13th November. Owing to the shallowness of the river she was not able to proceed further. She is presumed to be the first steamboat ever to navigate the Rhine. She returned to the Thames in 1818, where Watt made a number of experiments with her, and was finally sold to Danish owners, who ran her between Copenhagen and Kiel. Occasionally the ship made excursions in the Sound, at times it is recorded with the Danish royal family on board.

Majestic. 1816.

The Majestic was built during 1816 at Ramsgate by Court and engined by Dixon. She was a vessel of 90 tons burthen with engines of 24 n.h.p. employed on the Margate-London service. The Majestic did not limit her activities to passenger traffic, as it is related that on 28th August, 1816 she towed the East Indiaman Hope from Deptford to Woolwich at 3 miles an hour against the wind, an achievement regarded at the time as remarkable.

Britannia. 1817.

The Britannia was built in 1817 by Titterton at Stockwith near Gainsborough. She was a vessel of 70 tons burthen and has 15 n.p.h. engines by the Butterly Company. She worked between London and Southend.

Sons of Commerce. 1817.

This vessel was constructed in 1817* at Blackfriars by Lafort and engined by Boulton & Watt to the order of Elias Evans, a shipwright of Frindsbury, Rochester. She was 85 feet in length by 25.25 feet beam, 8 feet in depth and of 77 tons burthen. The engines were of 20 n.h.p. having cylinders 26.75 in. by 30 in. The paddle wheels were 10 feet in diameter. The Sons of Commerce ran alternatively with the Thames on the Gravesend service, and was in service as late as 1828.

On 1st May, 1818, a licence was granted to a Joseph Jones to act as master or chief steersman of this vessel. He was probably the first master to be licensed to navigate a steamship.

Favourite. 1817.

The Favourite was built in 1817 for the Gravesend Company at Lafort's shipyard, which was close to Blackfriars Bridge. She was built for the Margate service and was a fairly large vessel of 160 tons burthen, having engines of 40 n.h.p. by Boulton & Watt. She ran on the Margate service for some years, was one of the fastest steamboats on the river and her name betokened her popularity. The licence was for 320 passengers. In the return of river steamers in commission her name appears in the 1828 list.

London. 1817.

The London, built in 1817 by Searle, of Westminster and engined by the Butterly Co., was a vessel of 70 tons

* See list of Steamboats, pages 18 and 19.

burthen and 14 n.h.p. She was 85 feet in length by 14 feet beam. The engine had only one cylinder and was of the "grasshopper" type. The boiler was also of a curious design, being very narrow and was on either side of the engine. As late as 1830 the LONDON was running from Richmond to Gravesend, Southend and Sheerness.

DIANA. 1817.

The DIANA was built to run in conjunction with the LONDON. Built in 1817,* by Brent at Rotherhithe and engined by Boulton & Watt she was in service as late as 1828.

VICTORY. 1818.

The VICTORY was built under the direction of a Mr. Dodd, C.E., in 1818, by Evans, at Rotherhithe. She was a vessel of 160 tons burthen with engines of 40 n.h.p. by Dixon, and was intended for service between Richmond, Gravesend and Margate. Her first owners failed, but she was made to pay later and ran very many years on the Margate and Thames excursion service. In 1842 she collided with a collier brig off Barking when she had 400 passengers on board, luckily she did not sink.

LONDON ENGINEER. 1818.

A number of experimental steamers have been constructed from time to time which might well be described as freaks. Of these the LONDON ENGINEER was one of the most prominent examples. She was built in

* The Waterman's History gives the year as 1821.

1818 for the Margate service by Brent of Rotherhithe and engined by Maudslay of Lambeth.

Her dimensions were 120 feet long, beam 24 feet, with a draught of 5 feet, tonnage 315. The machinery was of a peculiar design, being of the bell-crank pattern, having two vertical cylinders 36 inches diameter with a 30 inch stroke. The cylinders were placed on either side of the vessel driving a pair of internal paddle wheels. The two paddles made 28 revolutions a minute, and the stream of water to them was kept constant by air forced into the waterway by two large pumps. The boiler pressure was 5 lb. per sq. inch and steam was supplied by three single-furnace copper boilers.

As the whole contraption was not in any way a success it was not surprising that the LONDON ENGINEER did not have a long life. She was a fine vessel in her way, having a nice saloon aft, with the luxury of upholstered settees, although the passengers in the cheaper fore-cabin had to be content with wooden benches. She was too big a vessel to have her sail set on the funnel and was given two proper masts, although the funnel with its decorated top was nearly as tall as the masts.

The author has no record of any steam vessel being built in the interval between the LONDON ENGINEER and the RAPID. When once the certainty and safety of the river steam packets was established, it was not surprising that they gained immense popularity, not only as a means of transport pure and simple, but also for excursion purposes. It is related on good authority that from 1820 to 1824 the number of passengers exceeded a million

REFERENCES AND AUTHORITIES OF THE EARLY STEAMBOATS.

		Name	Builders	Engines by	Service
A	1813				Gravesend—London
S	1814				From Bath
W	1813	MARGERY (a)	J. & C. Wood, Glasgow	Robertson	London—Gravesend
S	1813	MARGERY	J. & C. Wood, Glasgow	?	London—Gravesend
R	1814	MARGERY	J. & C. Wood, Glasgow	?	London—Gravesend
A	1814	MARGERY	J. & C. Wood, Glasgow	?	London—Margate
W	1813	ARGYLE (b)	J. & C. Wood, Glasgow	Cook	London—Gravesend
S	1815	ARGYLE	J. & C. Wood, Glasgow		London—Margate
R	1816	ARGYLE	J. & C. Wood, Glasgow		London—Margate
A	1815	DEFIANCE			London—Margate
W	1815	HOPE	Laurence, Bristol	Johnson	London—Sheerness—Chatham
AW	1816	REGENT	Courthope, Rotherhithe	Maudslay	London—Margate
AS	1816	CALEDONIA	J. & C. Wood, Glasgow	Boulton & Watt	London—Margate
AW	1816	MAJESTIC	Court, Ramsgate	Dixon	London—Margate
W	1817	BRITANNIA	Titterton, Stockwith	Butterley	London—Southend
W	1817	SONS OF COMMERCE	Lafort, Blackfriars	Butterley	London—Gravesend
SAR	1817	SONS OF COMMERCE	Evans, Rochester	Boulton & Watt	London—Margate
AS	1817	FAVOURITE	Lafort, Blackfriars	Boulton & Watt	London—Margate
W	1818	FAVOURITE	Lafort, Blackfriars	Boulton & Watt	London—Margate
ASW	1817	LONDON	Searle, Westminster	Butterley	London—Richmond—Southend and Sheerness
S	1817	DIANA	Brent, Rotherhithe	Boulton & Watt	do.
W	1821	DIANA	Brent, Rotherhithe	Boulton & Watt	do.
AS	1818	VICTORY	Evans, Rotherhithe	Dixon	London—Margate
W	1819	VICTORY	Evans, Rotherhithe	Dixon	London—Margate
W	1814	RICHMOND	Lepingwell, Yarmouth	Maudslay	London—Richmond

S	1818	RICHMOND	Lepingwell, Yarmouth	Maudslay	London—Richmond
ASW	1818	LONDON ENGINEER	Brent, Rotherhithe	Maudslay	London—Margate
A	1820	RAPID	Greenock		London—Margate
W	1820	ECLIPSE	Brent, Rotherhithe	Boulton & Watt	London—Margate
A	1820	EAGLE	Deptford	Boulton & Watt	London—Margate
W	1821	EAGLE	Brocklebank, Deptford	Boulton & Watt	London—Margate
W	1821	VENUS	Brent, Rotherhithe	Boulton & Watt	London—Margate
W	1821	ALBION	Evans, Rotherhithe	Boulton & Watt	London—Margate
W	1821	SWIFTSURE	Wallis, Blackwall	Boulton & Watt	London—Gravesend
W	1821	HERO	Bankham, Medway	Boulton & Watt	London—Margate
W	1822	SIR J. YORKE	Joliffe, Chester	Butterley	Southend—Sheerness
AW	1822	ROYAL SOVEREIGN	Brocklebank, Deptford	Butterley	London—Ramsgate
W	1823	DART	Evans, Rotherhithe	Boulton & Watt	London—Margate
AW	1824	CITY OF LONDON	Brocklebank, Deptford	Boulton & Watt	London—Margate
A	1825	DUKE OF SUSSEX	Rotherhithe		London—Boulogne
RW	1825	HARLEQUIN	Evenden, Deptford	Butterley	London—Margate
W	1825	COLUMBINE	Evenden, Deptford	Butterley	London—Margate
AW	1826	MAGNET	Tebbutt, Limehouse	Boulton & Watt	London—Margate
W	1826	HAWK	Wallis, Blackwall	Boulton & Watt	London—Gravesend
W	1828	RAMONA	Evans, Rotherhithe	Butterley	London—Margate
SR	1828	SOPHIA JANE	Barnes Miller	Barnes, Miller	London—Gravesend
W	1828	SOPHIA JANE	Evans, Rotherhithe	Barnes, Miller	London—Gravesend
W	1828	ENDEAVOUR	Searle, Westminster	Maudslay	London—Ramsgate

(A) Aylmer. Early London River Steamers. "The Engineer," 9/9/1932.
(a) On the authority of W. Denny Bros. Ltd., built by William Denny and the name spelt MARJORY.
(b) Re-named THAMES.
(S) Sandham Proc. Inst. Mech. Engs., March, 1885. Paddle Wheel Steam Navigation.
(R) Ravenhill Proc. Inst. Naval Architects, 1880. Passenger Steamers on the Long Ferry.
(W) History of Waterman's Co., Vol. 3, 1874.

each year, and in one particular year exceeded 1,250,000, and the number of packets increased with equal rapidity.

RAPID. 1820.

Built 1820 for the Margate service, at Greenock. Powerfully engined for her size, and had a reputation for speed. Dimensions, 94 ft. by 17.3 ft. by 9 ft. Tons 88.

ECLIPSE. 1820. See G.S.N. Co., Chapter IV.

EAGLE. 1820. See G.S.N. Co., Chapter IV.

VENUS. 1821. See G.S.N. Co., Chapter IV.

ALBION. 1821.

Built by Evans, Rotherhithe, 1821. Engines by Boulton & Watt. London and Margate service.

SWIFTSURE. 1821.

Built by Wallis at Blackwall, 1821. Engines by Boulton & Watt. N.H.P. 30. Tons 105. London and Gravesend service.

HERO. 1821. See G.S.N. Co., Chapter IV.

SIR J. YORKE. 1822.

Built by Joliffe, Chester, 1822. Engines by Butterly Co. Southend and Sheerness service.

ROYAL SOVEREIGN. 1822. See G.S.N. Co., Chapter IV.

DART. 1823.

Built by Evans, Rotherhithe, 1823. Engines by Boulton & Watt. London and Margate service.

CITY OF LONDON. 1824. See G.S.N. Co., Chapter IV.

DUKE OF SUSSEX. 1825.

The DUKE OF SUSSEX was built in 1825 at Rotherhithe, with dimensions of 143.8 ft. length, 16.4 ft. beam and 4.2 ft. depth, giving a net tonnage of 105. She was quite a speedy little vessel, having engines of 80 n.h.p. Although the Admiralty did not consider that she was of sufficiently strong construction to carry the long 18 pounders which were usually reserved for the boats of the G.S.N. Co., they considered that she was quite big enough to have 6 guns on the broadside, either 6 or 12 pounders.

The DUKE OF SUSSEX usually ran on the excursion service between London and Boulogne until about 1850.

HARLEQUIN. 1825. See G.S.N. Co., Chapter IV.

The HARLEQUIN originally worked between Richmond and Margate, but was later transferred to the service between the Customs House and Margate—Ramsgate. On 9th April, 1830, she went on an excursion trip to the Nore and back with 400 passengers on board. In the 1840's she ran to Boulogne. Broken up about 1856.

COLUMBINE. 1825. See G.S.N. Co., Chapter IV.

The COLUMBINE was at first on the same service as the HARLEQUIN, but in the 1840's was running to Havre. In 1855 she stranded near Rotterdam.

MAGNET. 1826. See G.S.N. Co., Chapter IV.

The MAGNET worked between London—Ramsgate and London—Boulogne until 1847.

HAWK. 1826.

The HAWK was built by Wallis, of Blackwall, in 1826 for the Gravesend Steamboat Co., and engined by Boulton & Watt. Her dimensions were 105.6 feet by 17.5 feet by 5.3 feet and her tonnage 104 with engines of 50 n.p.h. For nearly 25 years she was in service between London, Gravesend and Sheerness.

SOPHIA JANE. 1826.

There is conflicting evidence regarding the date and builder of the SOPHIA JANE. Sandham and Ravenhill give the year as 1828 and the builders as Barnes & Miller, Steam Engine Manufacturers, off Glass House Fields, Ratcliffe. The Waterman's History gives the same year, but built by Evans, Rotherhithe. In "Pacific Steamers" by W. Lawson, reference is made to the SOPHIA JANE, and that writer gives the year as 1826, and the builders as Barnes & Miller. These last particulars have been accepted as correct.

The SOPHIA JANE was built to run between London, Gravesend and Margate and did so for a time. In dimensions she was 126 feet by 20 feet and of 256 tons burthen with engines of 50 n.p.h. She ran later between London and Calais and was subsequently sold to the St. George Steam Packet Co., Liverpool and ran between Liverpool and Douglas.

Her greatest achievement was when she steamed out to Australia, after being purchased by a Sydney firm. She arrived at Sydney on 16th May, 1831, calling at Pernambuco and the Cape on the way out. She ran on the Australian coastal traffic until 1845, when she was broken up and her engines put into the PHŒNIX, a vessel which had been built for them. The SOPHIA JANE was Australia's first steamship.

RAMONA. 1828. See G.S.N. Co., Chapter IV.

ENDEAVOUR. 1828.

The ENDEAVOUR, built by Searle, Westminster, and engined by Maudslay, worked between London and Margate.

ALBION. 1833.

An advertisement dated 18th June, 1833, states that the steamboat ALBION has been chartered by T. Jones in conjunction with J. Winsor to run a trip from London to the Nore. It informed the public that: " The company are desired to assemble at the New Pier, St. Katherine's Dock, where they will embark Gratis, at a Quarter before Eight, a.m. precisely, as the utmost punctuality will be observed." The advertisement further goes on to say that the promoters " trust that the arrangements they have made for their convenience and conviviality, will meet with the approbation it will be their happiness to merit. The vessel will proceed along the Kent and Essex shores; and from the Deck will be visible the various works of art and nature that grace those romantic situations; the Thames Tunnel, that effort of national science and industry, which if completed would be the glory of the world; that splendid Palace for decayed Seamen GREENWICH HOSPITAL; WOOLWICH DOCKYARD; GRAVESEND PIER and FISHERIES; and TILBURY FORT, from whence Queen Elizabeth addressed her patriotic partisans, to the ultimate destruction of the Spanish Armada; and so onward to the Nore. The Company will also have an opportunity of witnessing THE LAUNCH OF THE WATERLOO 120 GUNS."

The company was also to be entertained with a concert, a wonderful illusionist, a ball, and an efficient band. The price of the tickets was 5/- each, including pier dues.

GREENWICH. 1835.

The GREENWICH was built at Poplar in 1835, for the Greenwich Steam Packet Co., a typical river paddle steamer of 96 tons net. Her dimensions were 122.3 ft. by 15.1 ft., while her engines were of 40 n.p.h. The Admiralty considered she was strong enough to mount a long 18 pounder gun in case of war.

She was employed on the passenger service between Greenwich and London Bridge Wharf and is principally remembered by an unlucky collision in October, 1836. She ran into the ROYAL TAR of Dublin and was in the gravest danger of sinking with a large number of passengers on board. A minor panic developed until the Milton and Gravesend Company's steamboat FLY came alongside and took them off.

ECLIPSE, ISLE OF THANET, FAWN. 1839.

In 1839 and 1840 David Napier built these three steamers fitted with high pressure engines specially for the traffic to Margate, Ramsgate, Deal and Dover. The general public considered them highly dangerous boats, the ECLIPSE which like the ISLE OF THANET was double funnelled being called the "Death and Glory boat." This was the steamer immortalised in "the Ingoldsby Legends," published in 1840.

"If in one of the trips of the steamboat Eclipse
You should go down to Margate to look at
the ships"

LOCOMOTIVE. 1840.

The LOCOMOTIVE built about 1840, by Ditchburn & Mare and engined by Braithwaite & Milner entirely on the plan of a locomotive high pressure boiler and engine. The exhaust was carried direct into the boiler funnel, causing similar beats to a locomotive; hence the boat's name. She worked between London and Richmond and Hampton Court.

CARDINAL WOLSEY. 1840.

Built for same service as the LOCOMOTIVE, and fitted with a single direct-acting engine by the same makers.

SONS OF THE THAMES and FATHER THAMES. 1844.

These were both built in 1844 for the Thames Steamboat Company. They ran between Hungerford Market and Gravesend. The SONS OF THE THAMES was of the following dimensions: 140 ft. by 17 ft. by 8.3 ft., tonnage 199, n.h.p. 56. The FATHER THAMES was larger, 141.6 ft. by 19.2 ft. by 10 ft., tonnage 254, n.h.p. 70. They were both withdrawn in November, 1845, in view of the opening of the North Kent Railway to Gravesend. The FATHER THAMES was taken over by the Diamond Steam Packet Company and continued in service until about the middle 1850's.

EAGLE & FALCON. 1845.

The EAGLE and FALCON were two steamboats run by the Eagle Company, a rival of the Thames Steamboat Co. They also started from Hungerford Pier making two trips a day against the other Company's three.

GEMINI. 1850.

The GEMINI was a freak. She was a " safety " iron twin hulled steamer, designed by a Mr. Peter Borrie and launched by him in the summer of 1850. Her dimensions were no less than 157.6 ft. in length by 26.6 ft., each hull being 8.5 ft. wide with a gap of 9.6 ft. between. The single paddle wheel worked in this gap, and she was intended to carry goods, passengers, cattle and all types of vehicles. It was expected that there would be a huge passenger traffic on the Thames during the 1851 London Exhibition, and it was hoped that the GEMINI would eclipse in speed and dimensions the large fleet of steamboats plying between London Bridge and Gravesend, some of which were remarkably speedy. The shape of

the twin hulls afforded a large amount of deck-room with a very small tonnage, and considerably decreased the resistance to the water. The bow and stern were similar and she was expected to steam with equal facility ahead or astern, being fitted with rudders fore and aft.

According to the standard of the day, there was ample accommodation for between* 800 to 1,000 passengers on board, and altogether she cost her inventor some £14,000. Unfortunately for him, on her trial trip the speed was practically nil, and she found it impossible even to stem the tide. The trial was the only trip she ever made and in consequence she never plied for hire.

THUNDER, LIGHTNING, ECLIPSE, THISTLE, MINERVA, ERA and VIVID.

An account of the river traffic in 1842 gives particulars of seven above-bridge steamboats; the length of these varied between 80 and 100 feet, n.h.p. 22-27. Speed ranged from 8.5—10 knots. They were named THUNDER, LIGHTING, ECLIPSE, THISTLE, MINERVA, ERA and VIVID.

* * * *

It will be noticed in these pages what an important place Margate was and still is in the down river passenger traffic. Margate made a success of many services and the many steamboat services made a success of Margate even before the advent of the railway in 1846. In the century

* Certified by the Board of Trade to carry 916 persons.

that has passed since then, improved and faster steamboats together with cheap facilities offered by two very active competing railways (until they amalgamated in 1898) trebled the population of the place. *The Illustrated London News*, in which so many interesting phases of London life are mirrored, said in 1844 " London Bridge Wharf is the main point for the Margate steamboats. It is curious to turn to the statistics of steam navigation to this watering place. It appears that the first steamboat the THAMES, 90 tons, left London for Margate in 1815. In 1817 the fares were 11/- and 15/- each person. The passage was made in 10 or 12 hours whereas it is now effected with ease in less than half that time. The old sailing packets were occasionally 30 and even 36 hours on the same passage and such was the popular apprehension of explosion on board the first-built Margate steamboats that we remember a scientific friend addressing a long letter to *The Times* to prove their safety. Margate almost within the present century was a poor inconsiderable fishing town built for the most part in the valley adjoining the harbour, the houses of which were in general mean and low. One dirty narrow lane called King Street being the principal street of it. The shore is well adapted to sea bathing and to this fact and the facility of communication with the metropolis by means of steam vessels must be attributed the rapid increase in the population which now (1844) amounts to 11,000." The population is now, in 1949, given as 40,390.

CHAPTER II.

PERIOD 1829 to 1876.

BETWEEN the years 1829 and 1846, no less than fourteen steamboat companies (apart from the G.S.N. Co.) put craft in commission on the London river, three of which confined their services to the water above London Bridge.

The Companies concerned were:

The Milton & Gravesend Steam Packet Company.
The Margate & London Steam Packet Company.
The Star Steam Packet Company.
The Gravesend Steam Packet Company.
The London & Westminster Steamboat Company.
The London & Herne Bay Steam Packet Company.
The Woolwich Steam Packet Company.
The Diamond Steam Packet Company.
The Iron Steamboat Company.
The Waterman's Steam Packet Company.
The Blackwall Railway Company.
The Thames Steamboat Company.
The Citizen Steamboat Company.
and The Halfpenny Fare Steamers.

THE MILTON & GRAVESEND STEAM PACKET COMPANY.

This Company was formed in 1829, and their first steamboat, the KENT, was launched on the 29th April, 1829. This boat was followed soon afterwards by the

PEARL and the ESSEX, and in 1830 by the FLY. These boats ran between Gravesend and London. The ESSEX and KENT were sold to the Diamond Company in 1835.

Date	Name	Builder	Tons	Engines by	H.P.
1829	KENT	Baukham, Gravesend	135	Seaward	80
1829	PEARL	Fletcher, Fearnall, Limehouse	146	Seaward	70
1829	ESSEX	Fletcher, Fearnall. Limehouse	132	Seaward	70
1830	FLY	Wallis, Blackwall	95		25

THE MARGATE & LONDON STEAM PACKET COMPANY.

During the late 1820's a steam packet company was formed with headquarters at Margate. It was officially designated The Margate & London Steam Packet Company, but was better known as "The Royal Line." The service commenced in 1830. The Company's first vessel was the WILLIAM IV, followed by the ROYAL GEORGE, ROYAL ADELAIDE and the ROYAL WILLIAM.

In 1835, the Company was taken over by the G.S.N. Company, but the names do not appear in that Company's list of vessels.

Date	Name	Builder	Tons	Engines by	H.P.
1830	WILLIAM IV	Evans, Rotherhithe	105	Boulton & Watt	50
1830	ROYAL GEORGE	Fletcher & Fearnall	240	Seaward	140
1830	ROYAL ADELAIDE	Snook, Limehouse	178	Seaward	120
1831	ROYAL WILLIAM	Snook, Limehouse	233	Seaward	160

THE STAR STEAM PACKET COMPANY.

The Star Steam Packet Company was founded in 1833, and operated a service of steamboats between London and the Town Pier, Gravesend, the service opening on 29th

Vessels of the Star Steam Packet Company.

Date	Name	Builders	Tons	Dimensions	Engines by	N.H.P.	Cyls. (ins.)	Speed (knots)
1834	Star	, Limehouse	300	141 × 21 × 11	?	120		
1834	Comet	?	163	135 × 16 × 9.4	?	70		
1834	Mercury	?	252	143.5 × 21 ×	?			
1836	Vesper	Fletcher & Fearnall	250	155 × 18 × 9.6	Miller, Ravenhill	90		
1836	Satellite	?	199	139.10 × 17 × 8.4	?	70		
1836	Planet	Rotherhithe	184		?			
1844	Meteor	Miller & Ravenhill	274	170 × 18 × 9	Miller, Ravenhill	80	37 × 36	13.25
1847	New Star	Miller & Ravenhill	271	180 × 17 × 9	Miller, Ravenhill	70	34 × 33	14.00
1850	Jupiter	Miller & Ravenhill	280	165 × 18 × 9	Miller, Ravenhill	80	37 × 36	14.00
1852	Mars	Money, Wigram & Sons	291	180 × 18 × 10	Miller, Ravenhill	90	39 × 36	14.00
1854	Venus	C. J. Mare & Co.	291	180 × 18 × 9.5	Miller, Ravenhill	90	39 × 36	14.75
?	Medway	?	?		?			

July, 1834, the first boat being the STAR which landed 500 passengers from London, and subsequently—in conjunction with the COMET—made an excursion down the river.

The METEOR was an iron vessel which had a beam engine that was originally in the PLANET, she also had fixed floats to the paddle wheels; the NEW STAR, JUPITER, MARS and VENUS had oscillating engines and feathering paddle wheels. The VESPER was built of wood.

The METEOR, in the autumn of 1854, whilst returning from Gravesend on a dark foggy night, came into collision with a large steamer which cut the METEOR completely in halves, although she was an iron vessel. Fortunately there was no loss of life.

After the dissolution of the Company, the MEDWAY was broken up. The COMET was sold and ran between Hartlepool and Sunderland, the VENUS went to the Channel Islands and ran between Guernsey and Jersey; the SATELLITE went to Liverpool; the JUPITER was purchased by the Dover Mail Steam Packet Co., who overhauled both hull and engines and also re-boilered her. On completion of the work, the Board of Trade surveyor granted a passenger certificate for the summer season, but the Admiralty mail packet surveyor refused to grant one; hence, while you might cross the Channel in the JUPITER, any letter you had posted could not accompany you. The JUPITER was ultimately turned into a blockade runner, and after a severe struggle in the Bay of Biscay, fell a victim to Neptune.

The G.S.N. Co. purchased the MERCURY in 1841 and the STAR in 1846.

Face page 32

P.S. " Royal Adelaide " (1830)

THE LONDON & CLACTON STEAM PACKET COMPANY.

The London & Clacton Steam Packet Company was established in 1834, by Mr. Jackson, an energetic and prominent citizen of Woolwich, who may be said to have been the founder of Clacton-on-Sea. To further the popularity of Clacton, he established the L. & C.S.P. Co. Unfortunately, the author can find no trace of any vessels which worked the service.

THE GRAVESEND STEAM PACKET COMPANY.

The Gravesend Steam Packet Company commenced a river service in 1835 between London, Gravesend, Sheerness, Margate and Ramsgate, and for several years secured their share of the increasing river traffic. The Company had four steamers; the ROSE built in 1832 and purchased in 1835; the ROYAL TAR and the GIPSEY, both built in 1835; and the WATER LILY in 1836; all built on the Thames, of 84 gross tons, and 40 n.h.p.

THE LONDON & WESTMINSTER STEAMBOAT COMPANY.

In 1835, the London & Westminster Steamboat Company instituted a non-stop service between the Old Swan Pier and an exclusive landing stage which then existed at the southern end of Westminster Bridge alongside the site of the present London County Hall. Morning and evening the boats were crowded; they were popularly known as the " flower boats," as all were named after flowers, the name being painted on the paddleboxes. They were all double-enders, and had oscillating engines by J. Penn & Son. The funnels were bell-topped, painted black with a white band near the top. The available particulars of these boats are as follows:

Date	Name	Built at	Dimensions (feet)	Tons	N.H.P.
	AZALEA				
	BLUEBELL				
	ROSE				
1845	DAHLIA	Blackwall	114.5 × 13.1 × 6.2	57	24
1847	LONDON PRIDE	Blackwall	101.1 × 13.7 × 5.6	72	24
1847	SUNFLOWER	Blackwall	101 × 13 × 6.5	70	24
1856	LOTUS	Deptford	105.2 × 13.7 × 5.6	65	24
1858	CAMELIA	Deptford	108.8 × 14.2 × 6.3	66	32

These boats had a speed of 10.5 knots, and were driven by paddle wheels of 8 feet diameter with fixed floats. In 1849, the paddles of the LONDON PRIDE were replaced by wheels of 9 ft. 4 in. diameter having feathering floats, the speed being increased thereby to 13 knots.

The DAHLIA, LONDON PRIDE, SUNFLOWER, LOTUS and CAMELIA had many changes of ownership, and eventually became part of the fleet of the London Steamboat Company.

According to the Mercantile Navy List the subsequent owners of these vessels are as follows:

DAHLIA	1871-1872	J. H. Wilson, London.
	1873-1875	Citizen Steamboat Co.
	1876-1881	London Steamboat Co.
	1882	H. J. Barker, London.
	1883-1884	London Steamboat Co.
	1885-1889	River Thames Steamboat Co.
	1890	Not recorded in the Mercantile Navy List.
LONDON PRIDE	1871-1874	R. W. Jennings, London.
	1875	W. H. Stratton, London.
	1876	G. A. Cape, London.
	1877-1884	London Steamboat Co.
	1885	Not recorded in the Mercantile Navy List.
LOTUS	1871-1873	J. H. Wilson, London.
	1874-1875	Citizen Steamboat Co.
	1876-1884	London Steamboat Co.
	1885-1889	River Thames Steamboat Co.

LOTUS (*continued*)	1890-1896	Victoria Steamboat Association.
	1897	Victoria Steamboats, Ltd. *Renamed* LOBELIA.
	1898-1908	Thames Steamboat Co.
	1909	Not recorded in the Mercantile Navy List.
SUNFLOWER	1871	R. W. Jennings, London.
	1872-1874	J. H. Wilson, London.
	1875	W. H. Stratton, London.
	1876-1885	London Steamboat Co.
	1885-1888	River Thames Steamboat Co.
	1889	Not recorded in the Mercantile Navy List.
CAMELIA	1871	R. W. Jennings, London.
	1872-1875	Woolwich Steam Packet Co.
	1876-1884	London Steamboat Co.
	1885-1894	River Thames Steamboat Co.
	1895	Not recorded in the Mercantile Navy List.

THE LONDON & HERNE BAY STEAM PACKET COMPANY.

About 1835, the London & Herne Bay Steam Packet Company was formed to run steamers between London, Herne Bay and Margate. Their two best boats were the CITY OF CANTERBURY and the RED ROVER, said to be two of the fastest on the Thames. Another of their steamers,, the CITY OF BOULOGNE was eventually taken over by the South Eastern & Continental Steam Packet Company who ran her on the Folkestone-Boulogne service.

Date	Name	Built at	Tons	N.H.P.
1835	RED ROVER			
1835	CITY OF CANTERBURY	Poplar	260	120
1838	CITY OF BOULOGNE	Barking	270	120

RED ROVER.

In 1836 the RED ROVER was wrecked.* Presumably there was a second boat named RED ROVER, as it is recorded

* On the authority of N. W. Kennedy, *Records of Early British Steamships.*

that " the RED ROVER was always regarded as a fast boat; in 1843 she was considerably modernised, and later in that year was tried against the screw steamer MERMAID which had lately been fitted with Rennie's patent propeller. From the Long Reach Tavern down to Gravesend—a ten mile course—the MERMAID gained about 300 yards."

One Sunday afternoon in November, 1843, the RED ROVER left Herne Bay on her return to London with between 200 and 300 passengers and when close to the Nore and about five miles off shore, a loud noise was heard causing great consternation amongst the passengers. The captain gave orders to let go the anchor. After the confusion had somewhat died down it was ascertained that the piston crosshead had snapped, thus disabling one of the engines. The RED ROVER proceeded with only one engine, arriving at Blackwall three hours late.

HERNE.

In 1845, the steamer HERNE—probably owned by this company—running between Herne Bay and London, was lengthened 15 feet, giving increased accommodation for fore-cabin passengers. The alteration had the effect of increasing her speed. On her trial trip after the lengthening she ran from Blackwall to Herne Bay in 3 hours 49 minutes.

THE WOOLWICH STEAM PACKET COMPANY.

In 1834 was established a well-known and successful line of river steamers, the Woolwich Steam Packet Company, which eventually became the most important boat Company on the Thames and owned the largest fleet of steamers. This Company became famous by the introduction of Saloon Steamers.

The W.S.P. Co. started in a small way by the running of local services until they subsequently ran as far as the Thames Estuary. Their original services started from Charing Cross and ran to Strother's Wharf, High Street, Woolwich. The service commenced in 1834 with only two double trips daily, leaving Woolwich at 9.0 a.m. and 2.0 p.m. and returning from Charing Cross at 11.0 a.m. and 4.0 p.m., calling both ways at Greenwich and Queenhithe. The fares were very low, that for the main cabin being 1/- and for the second cabin 9d., with a Sunday all-round fare of 1/-. At one time these Woolwich boats were regarded as the dirtiest and most uncomfortable running on the London river.

As river traffic became more popular and the railways had not yet obtained the monopoly of the traffic on both sides of the river, the services were extended to Southend and Sheerness.

In 1865, the Woolwich Company amalgamated with the Iron, Citizen and Westminster Companies and in 1871 with the Waterman's Steam Packet Company and all their piers were brought into the Woolwich fold. This combined grouping secured for the Woolwich Company a monopoly of the Thames passenger traffic from Kew and Richmond up the river down to Southend and Sheerness at the mouth.

ALEXANDRA 1865/1875.

One of the steamers taken over was the ALEXANDRA, a handsome two-funnelled paddler having upper main deck saloons fore and aft of the paddleboxes. The ALEXANDRA was built in 1865 and originally intended as a blockade-runner, but the American civil war having ended before she was completed she was purchased by a concern known as The Saloon Steampacket Company, of London, and became the first large steamboat of the saloon type employed in passenger traffic on the river Thames between London Bridge and Gravesend. At the time she was advertised to carry 1,048 passengers which " can be so distributed that there need be no apprehension of overcrowding. She is substantially built, and her fitments and decorations are ingenious and tasteful without being unnecessarily costly, while the general arrangements for her management are such as will secure the favour of the public."

The ALEXANDRA was built by Kirkpatrick & McIntyre, of Port Glasgow, and fitted with diagonal oscillating engines by Smith & Co. of Greenock. The following advertisement appeared in *The Illustrated London News* of the date 30th September, 1865.

" GRAVESEND — THE GREAT EASTERN Steam-ship —
" The Saloon Steam-packet Company's vessel, the ALEX-
" ANDRA, leaves London Bridge Wharf Daily (Sundays
" included) at a Quarter past Ten a.m. for Rosherville and
" Terrace Pier, Gravesend, returning from Gravesend (calling
" at Rosherville) at Half-past Four o'Clock p.m.

Fares : From London to Gravesend :

" Fore Saloon (single)...1s. od. Grand Saloon (single)...1s. 6d.
" Ditto (return) 1s. 9d. Ditto (return) 2s. 6d.

From London round Great Eastern and Back :

" Fore Saloon 3s. od. Grand Saloon 4s. od.

Children under Ten years of age, half price.

" Offices, 45, Eastcheap."

In 1875, the Company purchased four steamboats from the Clyde; three of these had been built to re-establish a service between Helensburgh and the Gareloch by a company which had gone into liquidation.

DUKE OF CONNAUGHT, 1866/1875.

The first of these boats, the ARDENCAPLE was immediately renamed DUKE OF CONNAUGHT; she was built at Port Glasgow in 1866, by R. Duncan & Co. She had one boiler and two oscillating cylinders 28 in. diameter by 36 in. stroke of pistons.

DUKE OF CAMBRIDGE, 1866/1875.

The second steamer from the same Company was the ARDGOWAN, which was renamed DUKE OF CAMBRIDGE. Built in 1866 at Port Glasgow, by L. Hill & Co., she had oscillating engines and one haystack boiler.

DUKE OF TECK, 1866/1875.

The third vessel of the defunct Company was the LEVEN, which became the DUKE OF TECK. Built in 1866, by Blackwood and Gordon, Port Glasgow, she had oscillating engines and one haystack boiler.

DUKE OF EDINBURGH, 1870/1875.

Built as the CRAIGROWNIE, and renamed the DUKE OF EDINBURGH on coming to the Thames. Built in 1870 by R. Duncan & Co., Port Glasgow, to run in opposition to R. Campbell on the Kilmuir Station, but did not accomplish her object. For a short time she belonged to the Glasgow & South Western Railway. She had oscillating cylinders 28 in. by 44 in., and one haystack boiler.

FLEET OF WOOLWICH STEAM PACKET COMPANY
(as far as can be ascertained, excluding amalgamations).

Name	Built	Where built	Dimensions	Tons	N.H.P
SYLPH	1834				30
FAIRY	1834			77	
NAIAD	1835	Blackwall		76	36
ARIEL	1835		125 × 14.5 ×	85	
NYMPH	1836		123.6 × 14.5 × 6.9	88	36
DRYAD	1836		123 × 14 × 6.6	88	40
METIS	1836	Rotherhithe	141 × 16.3 × 6.8	94	40
OREAD	1836		151.5 × 14.9 × 7.2	103	50
DORIS	1836	Rotherhithe	130.4 × 15.7 × 6.6	78	40
ELPHIN	1836	Rotherhithe	148.2 × 13.9 × 7.9	109	50
WITCH	1844				
*ARIEL	1844	Rotherhithe	125.1 × 14.4 × 7.4	89	40
SIBYL	1844	Blackwall	114.5 × 13.5 × 6.9	82	32
*NYMPH	1852	Rotherhithe	123.6 × 14.5 × 6.9	88	36
PETREL	1853	Woolwich	152 × 16 × 7.7	118	50
*QUEEN OF THE ISLES	1853	Rochester	111 × 13 × 6.8	65	32
SEA SWALLOW	1854	Woolwich	153.7 × 15.6 × 7.6	128	60
OREAD	1854	Rotherhithe	151.5 × 14.9 × 7.2	103	50
*METIS	1857	Rotherhithe	141 × 16.3 × 6.8	94	40
CAMELIA	1858	Deptford	108.8 × 14.2 × 6.3	66	32
PLOVER	1858	Woolwich	125.9 × 14.2 × 6.5	63	32
QUEEN	1859	Millwall	120.5 × 13.5 × 6.8	69	32
*DORIS	1859	Woolwich	130.4 × 15.7 × 6.6	78	40
*FAIRY	1859	Woolwich	125.6 × 15.2 × 6.6	74	32
MARTIN	1860	Woolwich	116.8 × 14.2 × 6.9	65	32
QUEEN OF THE THAMES	1861	Woolwich	158 × 19 × 8.2	143	80
QUEEN OF THE ORWELL	1862	Govan	171.6 × 18.5 × 7.9	165	54
*NAIAD	1862	Woolwich	130.3 × 14.5 × 6.4	77	36
*NORTH KENT	†1863	Rochester	94.5 × 12.6 × 6.5	46	24
HERON	1863	Woolwich	117.4 × 14.2 × 6.5	64	36
¶ALBERT EDWARD	1865	Greenock	219.4 × 22.2 × 8.4	252	140
¶PRINCESS ALICE	1865	Greenock	219.4 × 22.2 × 8.4	252	140
ALEXANDRA	†1865	Port Glasgow	230.5 × 22.2 × 8.5	279	140
DUKE OF CAMBRIDGE	†1866	Port Glasgow	150.8 × 16.1 × 6.2	92	50

* Built of wood.
† Acquired in 1875.
¶ Acquired in 1866.

Name	Built	Where built	Dimensions	Tons	N.H.P.
Duke of Connaught	†1866	Port Glasgow	159.1 × 16.1 × 6.4	127	50
Duke of Teck	†1866	Port Glasgow	150.2 × 16.2 × 6.2	93	50
Prince Alfred	†1866	Nine Elms	107 × 13.9 × 6.3	59	20
Prince Arthur	†1866	Nine Elms	110 × 13 × 6.3	55	20
Cupid	1867	Blackwall	122.5 × 15 × 6.6	86	32
Vesta	1867	Blackwall	122.5 × 15 × 6.6	74	32
Duke of Edinburgh	†1870	Port Glasgow	175 × 17.1 × 6.8	123	70
Prince of Wales	1875	Battersea	110.4 × 13.9 × 6.4	64	24

On 7th March, 1863, at the height of the excitement caused by the entrance of Princess Alexandra into London, the Sibyl was completely wrecked at London Bridge. In passing the bridge, she struck something and immediately began to fill. The passengers, between 200 and 300 in number, were got off in safety, and shortly afterwards she sank and became a total wreck.

The Queen of the Thames and the Queen of the Orwell worked between London and Ipswich. In June, 1865, the former vessel ran down the ferry steamer to North Greenwich off Greenwich Pier. The ferry sank in two minutes, but a brig close at hand saved the passengers.

The Fairy had 25 in. cylinders with a 30 in. stroke, and the Plover had oscillating engines by Penn with 25 in. cylinders having a 27 in. stroke.

In 1875, the whole fleet became the property of the London Steamboat Company, and in the Mercantile Navy List of 1876 is shown under the new owners.

The Queen does not appear to have worked after 1875.

The Diamond Steam Packet Company.

The Diamond Steam Packet Company opened a London-Gravesend service in 1836, and ran in keen competition with the Star and the Woolwich Companies. The down river trips to Gravesend—the long ferry—used to leave London at 9.0 a.m., 10.30 a.m., 2.0 p.m., 3.0 p.m., 5.0 p.m. and 7.0 p.m. on weekdays, the single fares being 1s. 0d. for the fore cabin and 1s. 6d. for the chief cabin. On Sundays, the boats left at 7.0 a.m., 9.0 a.m., 10.30 a.m., 2.0 p.m. and 7.0 p.m., no distinction being made as to class, the single fare being 1s. 6d.

The Diamond steamboats carried a fore-mast which, like the funnel, could be lowered when negotiating London Bridge. A canvas awning was provided over the poop-deck—first class only—abaft the paddles.

A. R. Bennett in his book *London and Londoners in the Eighteen-fifties and sixties* says " their funnels, which had mostly straight tops, were painted black for a foot or so down, and then in four lines of diamonds to the base, white diamonds sideways and black diamonds foreward and aft." An old water colour painting of the BRILLIANT, in the possession of the author, shows the funnel with an ornamental top and, instead of the four rows of black and white diamonds, black and white squares—chessboard fashion.

The opening of the North Kent Railway to Gravesend did not seriously interfere with the river traffic, but the cheap fares subsequently introduced by the London, Tilbury and Southend Railway, together with the facilities and attractions offered by other Railway Companies, eventually so reduced the traffic that the services were totally discontinued in 1855, and the fleet sold and dispersed in 1856. The boats of the Star Company had previously been withdrawn, as also had those of the Thames Company and other boats running to Hungerford Market.

The following appeared in *The Illustrated London News* of 20th December, 1845. " Serious collision between the Emerald, Gravesend steamer, and the John Bull, Hamburg packet. At 10 minutes to 5 o'clock on Saturday evening, as the Emerald steamer was proceeding on her way from London to Gravesend, her captain observed a large steam packet directly ahead of him coming up the river. As she approached, he and his men in the bow of the Emerald shouted out, but the steamer kept on her course. The order to port the helm was scarcely obeyed when the Emerald was struck on her larboard quarter, just before the paddle-box, by the bow of the other vessel, which proved to be the John Bull, a Hamburg packet. The paddle-box was stove in and the funnel was cast down with a tremendous crash, producing a shock throughout the vessel. The confusion and dismay that followed was quite indescribable as the boat was thought to be going down. The passengers were all saved by the John Bull and the Blackwall Railway boat Railway."

The Ruby and Sapphire were purchased in 1854 by the General Steam Navigation Company. The Emerald (1) was purchased by the Licensed Victualler's Company who ran her to Boulogne; the Emerald (2) was converted to a schooner. The remaining boats were broken up. The Ruby was the best known of the fleet and the fastest packet on the river until the Star Company brought out their

METEOR in 1844. The RUBY had side lever engines and paddle wheels 17.5 feet in diameter. In *Thames Passenger Steamers*, by Ravenhill, it is stated that the RUBY was the boat which was pitted against each new Star boat as they came out. She was a wooden boat of a light build, and made the quickest passage on record, with all things in her favour, from London Bridge to Gravesend Town Pier Light—which is 27 miles, mid-channel course—in 1 hour and 28 minutes. Details of the Diamond boats are very limited, but those available are as follows:

Name	Date	Built at	Dimensions	Tons O.M.	N.H.P.
RUBY	1836	Blackwall	160 ×19 ×9.4	280	80
SAPPHIRE			170.5×18.1×9.6	351	Atmospheric engines
EMERALD (1)	1835				
EMERALD (2)	*1834	Gravesend	155 ×19 ×		100
DIAMOND	1835	Blackwall	145 ×16.6×9.6	200	70
GEM	1836	Blackwall			
PEARL	.1828		103 ×17.6×9.4	152	70
TOPAZ	1837	Blackwall	135 ×16 ×9.6	163	70
BRILLIANT	1823	Poplar			
NEW PEARL	1835			135	

The dates and places of building are taken from Kennedy's *Records of Early British Steamships.* As previously stated, the Diamond Company purchased in 1835 the ESSEX and KENT from the Milton and Gravesend Steam Packet Company.

The following advertisement is taken from *The Times* of May 8th, 1847:

> EXCURSIONS. — Diamond Packets. — To Southend and Sheerness, from London-bridge-wharf. — The public are respectfully informed, that on and after Whit Sunday, May 23, one of the commodious and

* Purchased as the PRINCE ALBERT.

fast Diamond Packets will leave London-bridge-wharf daily at 9 o'clock, calling at the Tunnel, Greenwich, Woolwich, Erith, and Gravesend piers, for Southend and Sheerness, returning from Sheerness at half-past 4, and Southend at 5. — Diamond-office Gravesend, May 5.

HUNGERFORD MARKET STEAM PIER. — Fast and splendid steam packets leave this pier daily, Gravesend, 9; Woolwich, Blackwall, Greenwich, and Thames Tunnel, every hour, and half-past every hour; London and Westminster bridges every five minutes; Vauxhall, Battersea, and Chelsea, every quarter of an hour.

Two other daily steamer services were advertised. One, " by the fast and favourite packets SONS OF THE THAMES or PRINCESS ROYAL," gave a trip from Hungerford and London Bridge to Sheerness, Southend and Chatham. The other offered a combined railway and steamer service to Blackwall and then to Grays and Gravesend every hour or hour-and-a-half from 9 to 7. These services were augmented as the summer advanced. Hungerford Pier was at the foot of Hungerford suspension bridge, now remembered only by the name of a footbridge next to Charing Cross railway bridge.

THE IRON STEAMBOAT COMPANY.

In 1838, the Iron Steamboat Company started an up river service to Chelsea and Kew, charging twopence or more according to distance. Their boats were larger and faster than those of the other up-river services. They were chiefly built by Ditchburn and engined by Penn. The boats had all-black funnels, with openwork bell-mouth tops, and had " Iron Boat Company " painted on the outside of the paddle boxes. They carried peculiar names as the following list shows:

Name	Date	Dimensions (ft.)	Tons	N.H.P.
DAYLIGHT	1838			
MOONLIGHT	1838			
TWILIGHT	1838	82 × 13.5 × 6.8	51	24
STARLIGHT	1838	82 × 13.7 × 6.8	85	20
BRIDE	1839			
BRIDEGROOM	1839	93.5 × 13.6 × 6.9	121	24
BRIDESMAID	1839			
MATRIMONY	1842	95 × 13.4 × 6.4	54	24
WEDDING RING	1844	97 × 13.9 × 6.6	50	24

A contemporary account of the MATRIMONY describes her as of 36 tons (net), with fore and aft cabins, fitted with a pair of oscillating steam engines with 22 in. cylinders, 24 in. stroke, wrought iron tubular boiler, and paddle wheels with wood floats; the other boats of the Company are believed to have been similar.

The DAYLIGHT was the first steamboat on the Thames to be built of iron. The boats passed through several changes of ownership:

TWILIGHT.	1871-74	London, Westminster & Vauxhall Steam Packet Company.
	1875	W. H. Stratton, London.
	1876-84	London Steamboat Company.
	1885-88	River Thames Steamboat Company.
	1889	Victoria Steamboat Association.
	1890	Name removed from the Mercantile Navy List.
STARLIGHT.	1871-5	London, Westminster & Vauxhall Steam Packet Company.
	1876-	London Steamboat Company.
	1877	Name removed from M.N. List.

BRIDEGROOM.	1871-4	London, Westminster & Vauxhall Steam Packet Company.
	1875	W. H. Stratton, London.
	1876-7	London Steamboat Company.
	1878	Name removed from M.N. List.
MATRIMONY.	1871-74	London, Westminster & Vauxhall Steam Packet Company.
	1875	W. H. Stratton, London.
	1876-84	London Steamboat Company.
	1885	River Thames Steamboat Company.
	1886	L. J. Pearce, London.
	1887	Name removed from M.N. List.
WEDDING RING.	1871-4	London, Westminster & Vauxhall Steam Packet Company.
	1875	W. H. Stratton, London.
	1876	London Steamboat Company.
	1877-8	G. A. Cape, London.
	1879	Name removed from M.N. List.

MOONLIGHT, DAYLIGHT and BRIDESMAID. No record.

On 31st August, 1859, the BRIDE came to an unfortunate end near London Bridge when she fouled a sunken barge with such force that some of her plates were driven in. She began to fill rapidly and sank, but all the passengers were safely landed.

THE WATERMAN'S STEAM PACKET COMPANY.

The Waterman's Company, an old guild dating back to the 14th or 15th century, possessed the greatest possible authority in connection with the navigation of the river, and even to this day passenger river steamers must be navigated above Gravesend by a member of this body.

When the Woolwich Company was formed, bitter antagonism was shown by the Thames watermen, who claimed that their very ancient privileges and rights had been ruthlessly trampled on. At first they vented their spleen by such petty tricks as blocking the way of the packets by barges, but when they found such tactics unavailing the Waterman's Guild, which was very wealthy and powerful, decided to carry the war into the enemy's camp by forming the Waterman's Steam Packet Company, with a fleet of twelve fast packets which were put into service between 1840 and 1844. A fierce competition at once arose between the rival concerns. The packets were purposely timed to start together and racing was indulged in without any restraint. The pace became too hot to last, the Waterman's Company gave up the contest, and it was eventually—in 1865—bought up by the Woolwich Company. Even when the railways reached Woolwich—the G.E.R. to North Woolwich in 1847 and the S.E.R. in 1849—the popularity of the Woolwich packets did not lessen to any marked extent.

The Waterman's Company's boats were not named originally, but were numbered 1 to 12. After the taking-over by the Woolwich Steam Packet Co. the boats were named as follows: 1 CYGNET, 2 IBIS, 3 OSPREY, 4 CERES, 7 FALCON, 10 MERLIN, 12 SWIFT. There were five boats named TEAL, WIDGEON, PENGUIN, STORK and TERN. These names most probably fit in with numbers 5, 6, 8, 9 and 11, of which no details are available. In 1853, a thirteenth boat named ELFIN was acquired, which probably was never numbered as the superstitions connected with that number might have affected its popularity. The following are details of the boats:

No. etc.	Date	Where Built	Dimensions	Gross Tons	N.H.P.
1	1843	Blackwall	112.8 × 14.1 × 6.7	77	32
2	1844	Blackwall	129 × 14.6 × 6.6	89	40
3	1844	Blackwall	115.3 × 14.1 × 6.6	76	32
4	1844	Blackwall	116.9 × 14.4 × 6.8	81	32
5					
6					

Face page 48

P.S, " Brilliant " (1823)

P.S. " Citizen A " (1845)

No. etc.	Date	Where Built	Dimensions	Gross Tons	N.H.P.
7	1840	West Ham	110 × 14.2 × 6.4	73	32
8					
9		Blackwall			
10	1844	Blackwall	123.1 × 14.4 × 6.8	84	36
11					
12	1844	Blackwall	122 × 14.1 × 6.8	79	36
ELFIN	1853	Rotherhithe	148.2 × 13.9 × 7.9	109	50

WATERMAN NO. 10.

The career of WATERMAN No. 10 well nigh came to an end early one Saturday morning in April, 1844. According to the *Illustrated London News,* she had left Woolwich a few days earlier on a trial trip with coal for a five days consumption; she ran down to Margate in 4 hours 14 minutes, afterwards crossing over to Ostend against a strong headwind in 5 hours 12 minutes. On the Saturday she returned via Calais and raced back to Dover against the PRINCESS ALICE. In spite of the cross-Channel boat hoisting two sails to the wind, it only arrived at Dover two minutes ahead of the Waterman boat. No. 10 left Dover at noon on a pleasure trip arriving off Hastings about 2.30 p.m. She ran too close inshore and struck a dangerous reef known as the Castle Rocks. Passing over the obstruction, she badly injured her bottom. To save the vessel she was run on to the sands off the promenade, where she filled as the tide rose, and at high water was completely covered.

A quantity of empty barrels were slung round her shortly afterwards, by which means she was got further in and on the day following she was pumped out and had her leaks stopped and eventually floated off.

At the time of the accident several of the Waterman's Directors and Mr. John Penn, the celebrated marine engineer, were on board.

The following list gives the subsequent owners of the Waterman's Company boats:

CYGNET.	1876-84	London Steamboat Company.
	1885	River Thames Steamboat Company.
	1886	Name removed from Mercantile Navy List.
IBIS.	1876-84	London Steamboat Company.
	1885	River Thames Steamboat Company.
	1886	Name removed from Mercantile Navy List.
OSPREY.	1876-78	London Steamboat Company.
	1879	Name removed from Mercantile Navy List.
CERES.	1876-84	London Steamboat Company.
	1885-89	River Thames Steamboat Company.
	1890-86	Victoria Steamboat Association.
	1897	Victoria Steamboats Limited.
	1898	Thames Steamboat Company.
	1899	Name removed from Mercantile Navy List.
FALCON.	1876-84	London Steamboat Company.
	1885	Name removed from Mercantile Navy List.
MERLIN.	1876-84	London Steamboat Company.
	1885-88	River Thames Steamboat Company.
	1889	Name removed from Mercantile Navy List.
SWIFT.	1876-84	London Steamboat Company.
	1885-88	River Thames Steamboat Company.
	1889	Name removed from Mercantile Navy List.
ELFIN.	1876-78	London Steamboat Company.
	1879	Name removed from Mercantile Navy List.

Apart from the above the Waterman's Company purchased in 1867 two larger vessels, one of which will always be remembered with sorrow by those interested in the success of the Thames pleasure steamer. These vessels were:

ALBERT EDWARD and PRINCESS ALICE.

These two vessels were sister boats and were built in 1865 by Caird & Company, of Greenock, for the Wemyss Bay Railway Company and were named KYLES and BUTE respectively. After running a short time this railway abandoned ship-owning and sold both vessels to the Waterman's Company in 1867. When taken over by the Woolwich Company in 1871 the names were changed to ALBERT EDWARD and PRINCESS ALICE. They were handsome vessels with flush decks, two funnels fore and aft of the paddle boxes and one mast. Constructed of iron they had the following dimensions: 219.4 ft. by 20.2 ft. by 8.4 ft.; gross tons 171; oscillating cylinders 44 in. by 45 in.; N.H.P. 140. Steam was supplied by two haystack boilers working at 30 lbs. per square inch. They were licensed to carry 936 passengers each for river service in smooth water.

To increase the passenger accommodation the Waterman's Company erected on the flush deck a forward and aft saloon having a promenade deck on top. After the alteration the Board of Trade granted certificates permitting the steamers to carry 850 passengers to Gravesend and 460 to Sheerness in the summer. In April, 1870, the certificates were amended to allow 872 passengers to Gravesend and 462 passengers to Sheerness. In March, 1872, they were amended once again the complement being raised to 936 to Gravesend and 486 to Sheerness. The gross tonnage was increased from 171 to 250, but it greatly reduced the reserve strength of the already weak structure—the hull plates being only 3/16 inch thick. They were at the time the most popular passenger steamers on the Thames.

On one occasion in 1873, the PRINCESS ALICE was chartered to take the Shah of Persia on a tour of the London Docks.

The tragic end of the Princess Alice is the worst river disaster on record in the British Isles. It was on Tuesday, 3rd September, 1878, that the vessel left London Bridge at 11.0 a.m. for a day trip to Sheerness, having on board some 900 passengers. The return journey was commenced at 4.15 p.m. Gravesend was left behind soon after 6.0 p.m., and at about 8.0 p.m. she was in sight of Woolwich Arsenal and was booked to stop at North Woolwich Pier. The tide was running out fast and the Princess Alice kept close to the shore, thus avoiding the tide's force; and having rounded Tripcock Trees Point on the south Bank, she headed across the river to the next point on the north bank known as the Devil's House. When about half-way across the reach, in mid-stream, off the Beckton Gas Works and just below North Woolwich gardens, the collier Bywell Castle—1,326 tons gross—struck the Princess Alice near the starboard paddle-box cutting into the body of the vessel to a depth of fourteen feet. The engine room rapidly filled with many tons of water and the middle part thus weighted and partially severed sank and the foremost half of the vessel rising at a considerable angle, broke off and turned round and headed down river before it too sank. In a little over five minutes all was over. Nearly 700 lives were lost in the tragedy and the author was acquainted with a whole family from Croydon who perished. In spite of conflicting evidence, the subsequent enquiry made it clear that the fault in navigation rested entirely on the Princess Alice. Some years later the Bywell Castle was sunk in the North Sea.

This overwhelming disaster dealt a very severe blow to the Thames river excursion traffic and, with the increase in competing forms of land transport, it has never regained its former importance and volume.

The ALBERT EDWARD continued on the Thames till 1888, when she was withdrawn and scrapped.

THE BLACKWALL RAILWAY COMPANY.

When the Blackwall Railway Company—originally called the Commercial Railway—completed their line from the Minories to Blackwall in 1840 (the terminal station being on the edge of the river at Brunswick Wharf) they commenced to run a line of packets to Margate and Ramsgate. After running the steamers for a few seasons, the Railway Company withdrew them and entered into an agreement with the Star Steam Packet Company to run a service from Brunswick Wharf in connection with the train service to and from Fenchurch Street station. During the time this service operated, no better one existed. The express service, both in the morning and the afternoon, between London and Gravesend occupied, with the tide about 1½ hours—against the tide some 20 minutes longer.

The Star boats used on this service were the JUPITER, MARS and VENUS. Although the passage between Blackwall and Gravesend was comparatively short, the service was a trying one for the boilers. The vessels had to start with a good head of steam, generally kept just under blowing-off pressure by the use of the donkey, and the fires could not be eased at the end of the passage until the vessels were close alongside the piers; again the donkey was brought into requisition, with the fire-doors and smoke-box doors all thrown open to check, as far as possible, the steam from roaring off.

Name	Date	Built at	Dimension	G. Tons	Knots
PRINCE OF WAIES	1841	Deptford	180 × ×		14
RAILWAY	1842	Blackwall	145 × 19 × 10	258	12
BRUNSWICK	1842	Blackwall	158 × ×	258	14
BLACKWALL	1842				
QUEEN		Greenwich	160 × ×		17
COURIER					

The PRINCE OF WALES had an inverted beam engine of 136 N.H.P. by Miller & Ravenhill: RAILWAY had oscillating engines of 90 N.H.P. by Penn & Sons: BRUNSWICK had direct-acting engines of 90 N.H.P. by Seaward, and BLACKWALL a steeple engine of 90 N.H.P. by Miller & Ravenhill.

The engines of the PRINCE OF WALES were second-hand, having been taken from one of the old Margate steamers. She soon established herself as one of the fastest boats on the Thames, although she did not hold that position for long. In attempting to do so she made herself very unpopular, and there were constant complaints of her wash and on one occasion this resulted in the sinking of a little trading schooner with loss of life; in addition her reckless navigation resulted in a collision with the yacht MARINA in 1862.

The Blackwall Railway had a " freak " vessel, iron-built at Blackwall by Ditchburn and Mare. She was of 30 N.H.P. propelled in an entirely novel way. The vessel was designed especially for trade on the river, and the paddles consisted of single blades of iron dipping almost perpendicularly into the water and being forced backwards by an iron arm, making her action like a grasshopper. This novel gear drove the vessel at about 10 knots, and had the advantage that it created no wash, and therefore did no damage to passing vessels. The vessel was named PROPELLER.*

THE HALFPENNY FARE STEAMERS (DYER'S HALL COMPANY).

The Dyer's Hall Company ran vessels in 1846 from Dyer's Hall Wharf, near London Bridge to the Adelphi. The boats were of 15 N.H.P. the boiler consuming about 2½ cwt. of fuel per hour. The hulls were constructed with the fore and aft ends alike, so as to run in either direction without turning. They had two rudders, one at each end, the one not in use being locked. The engines had oscillating cylinders built by Joyce & Company, and are said to have been on the compound system working with H.P. steam from boilers by Montgomery. The boats were named ANT, BEE, CRICKET and FLY. The CRICKET was soon withdrawn, being found unsatisfactory, and replaced by the NEW CRICKET which was still less fortunate. On Friday, 27th August, 1847, whilst waiting near London Bridge at the busiest time of the day, the boiler of the NEW CRICKET unfortunately exploded. The NEW CRICKET had a compound engine.

These boats must have been among the first examples of the utilisation of the dual expansion engine for naviga-

* Sandham gives date of construction as 1845, and the vessel as working between Blackwall and Greenwich.

Aylmer, Parker & Bowen give the date as 1840 and working between Blackwall and Gravesend.

tion, and tragically unfortunate examples they were. The NEW CRICKET was replaced by a new boat named SUNSHINE.

The explosion of the NEW CRICKET's boiler with the consequent loss of life made the public avoid the ha'penny boats afterwards, and the Company soon went into liquidation. This was very unfortunate for the Company, whose boats were as good as any of the others; the tragedy occurred through the foolhardiness of the engineer, who had screwed down the safety-valves because he thought the engines were somewhat sluggish, and afterwards went off duty without warning his relief. Of the 150 passengers on board, 17 were killed and 60 severely injured.

The ha'penny steamboats were greatly patronised by the poorer working classes of the Metropolis.

THE CITIZEN STEAMBOAT COMPANY.

The Citizen Steamboat Company was started in 1846, with an up-river service to Chelsea and Kew, to run in competition with the Iron Steamboat Company. Later they ran to piers below London Bridge. The new Company had the largest fleet of steamboats on the Thames with the exception of the Woolwich Steam Packet Company, and was the last steamboat company to be formed on the London river until the formation of the London Steamboat Company in 1876. Furthermore, it retained its individuality until the advent of the London Steamboat Company.

The boats were built of iron, having flush decks with below-deck cabins fore and aft. They were painted black, had tall black funnels with two red bands a little way down from the top which was bell-mouthed with fancy work. The Citizen boats were identified alphabetically, as CITIZEN A, CITIZEN B, and so on, from A to S omitting I; on the paddleboxes were the City Arms with a large capital letter, and on the inner or deck side of the paddlebox they were named after various City Guilds, such as HABERDASHER, FISHMONGER, LORINER, MERCER, LEATHER-SELLER, SPECTACLE-MAKER, etc. A contemporary reference to CITIZEN E describes her as " flat bottomed." The last boat CITIZEN S was purchased from Mr. J. H. Wilson, of London, in 1873.

In 1876, the boats of the Citizen Company with those of the Iron Steamboat and the London & Westminster Companies became merged into the London Steamboat Company. A handbill dated September 1st, 1886, advertised CITIZEN S as running from London Bridge to Hampton Court and being " specially adapted for shallow water up the river." A special fare of one shilling was offered from any pier to Battersea Park pier " including admission to the Albert Palace."

Date	Letter	Built at	Dimensions	Tons	N.H.P.
1845	A	West Ham ...	94 × 13.4 × 6.6	61	24
1870	B	Battersea ...	109.6 × 13.8 × 6.3	61	24
1863	C	Battersea ...	105 × 13.8 × 6.2	61	24
1864	D	Battersea ...	105 × 13.8 × 6.2	61	24
1866	E	Battersea ...	121.2 × 14 × 6.2	77	24
1872	F	Battersea ...	119 × 13.8 × 6.4	73	24
1846	G	West Ham ...	94 × 13.4 × 6.6	61	24
1847	H	West Ham ...	113.4 × 14 × 6.2	58	24
1872	J	Battersea ...	127.6 × 13.7 × 6.5	77	24
1847	K	West Ham ...	103.5 × 12.3 × 6.4	75	24
1847	L	West Ham ...	103.5 × 12.3 × 6.4	75	24

Date	Letter	Built at	Dimensions	Tons	N.H.P.
1847	M	West Ham ...	105 × 13.5 × 6.2	75	24
1847	N	Blackfriars ...	101 × 11.9 × 6.3	62	24
1855	O	Battersea ...	88.5 × 12.7 × 6.2	57	24
1860	P	Battersea ...	108.6 × 13.5 × 6.4	58	24
1862	Q	Battersea ...	105 × 13.8 × 6.2	61	24
1865	R	Battersea ...	117 × 13.7 × 6.2	72	24
1866	S	Battersea ...	120.3 × 13.9 × 6.3	78	24

The foregoing particulars are taken from the Mercantile Navy List for the year 1871. Apparently B—F and J were replacements.

The list below gives subsequent owners of the Citizen boats:

CITIZEN A.	1876-84	London Steamboat Company.
	1885-89	River Thames Steamboat Company.
	1890-96	Victoria Steamboat Association.
	1897	Victoria Steamers Limited.
	1898-1908	Thames Steamboat Company.
	1909	Name removed from Mercantile Navy List.
		Re-named AZALEA in 1886.
CITIZEN B.	1876-84	London Steamboat Company.
	1885-89	River Thames Steamboat Company.
	1890-91	Victoria Steamboat Association.
	1892	Name removed from Mercantile Navy List.
CITIZEN C.	1876	G. A. Cape, London.
	1877-81	London Steamboat Company.
	1882	H. J. Barker, London.
	1883-84	London Steamboat Company.
	1885-89	River Thames Steamboat Company.
	1890-91	Victoria Steamboat Association.
	1892	Name removed from Mercantile Navy List.
		Re-named CARNATION in 1885.
CITIZEN D.	1876	G. A. Cape, London.
	1877-84	London Steamboat Company.
	1885-89	River Thames Steamboat Company.
	1890-96	Victoria Steamboat Association.
	1897	Victoria Steamers Limited.
	1898-1908	Thames Steamboat Company.
	1909	Name removed from Mercantile Navy List.
		Re-named DAISY in 1895.
CITIZEN E.	1876-84	London Steamboat Company.
	1885-88	River Thames Steamboat Company.
	1889	Name removed from Mercantile Navy List.
CITIZEN F.	1876-84	London Steamboat Company.
	1885-89	River Thames Steamboat Company.
	1890	Victoria Steamboat Association.
	1891	Name removed from Mercantile Navy List.
		Re-named FUCHSIA in 1885.

CITIZEN G.	1876-83	London Steamboat Company.
	1884	Name removed from Mercantile Navy List.
CITIZEN H.	1876	G. A. Cape, London.
	1877-84	London Steamboat Company.
	1885	Name removed from Mercantile Navy List.
CITIZEN J.	1876-84	London Steamboat Company.
	1885-89	River Thames Steamboat Company.
	1890-96	Victoria Steamboat Association.
	1897	Victoria Steamers Limited.
	1898-1908	Thames Steamboat Company.
	1909	Name removed from Mercantile Navy List. Re-named JESSAMINE in 1892.
CITIZEN K. } CITIZEN L. }	1876-83	London Steamboat Company.
	1884	Names removed from Mercantile Navy List.
CITIZEN M. } CITIZEN O. }	1873	Names removed from Mercantile Navy List.
CITIZEN N.	1876-84	London Steamboat Company.
	1885-88	River Thames Steamboat Company.
	1889	Name removed from Mercantile Navy List.
CITIZEN P.	1876-84	London Steamboat Company.
	1885	River Thames Steamboat Company.
	1886	Name removed from Mercantile Navy List.
CITIZEN Q.	1876-81	G. Dixon, Stockton-on-Tees.
	1882-98	Imperial Tramways Co., London.
	1899	Name removed from Mercantile Navy List.
CITIZEN R.	1876-84	London Steamboat Company.
	1885-89	River Thames Steamboat Company.
	1890-96	Victoria Steamboat Association.
	1897	Victoria Steamers Limited.
	1898-1908	Thames Steamboat Company.
	1909	Name removed from Mercantile Navy List. Re-named REDFERN in 1885.
CITIZEN S.	1876-84	London Steamboat Company.
	1885-89	River Thames Steamboat Company.
	1890-96	Victoria Steamboat Association.
	1897	Victoria Steamers Limited.
	1898-1908	Thames Steamboat Company.
	1909	Name removed from Mercantile Navy List. Re-named SNOWDROP in 1892.

THE THAMES STEAMBOAT COMPANY

This company owned the two steamboats mentioned on page 25.

CHAPTER III.

THE LONDON STEAMBOAT COMPANY AND ITS SUCCESSORS.

IN 1876, the London Steamboat Company came into being, and as previously stated took over the boats of the Woolwich, Citizen, and Waterman Companies, and a few other miscellaneous boats, making a grand total of 70: 30 from the Woolwich Company, 14 from the Citizen Company, 8 from the Waterman's Company and 18 miscellaneous.

The Company suffered a staggering blow when the PRINCESS ALICE was lost in 1878. This is referred to on page 52 and was fully described by the well-known authority Mr. F. G. Bowen in the *Engineer* of September 11th, 1942. It never recovered, and after working a few years the boats and plant were offered for sale by auction in lots in December, 1884. Some of the plant may have been sold to others, but the boats and chief assets were taken over by a new company—the River Thames Steamboat Company.

The new company was, itself, soon in difficulties. The London steamboats were a much used form of travel in those days and it would seem that the company should have been a solid success. The Thames is not the sweetest of rivers; still, during several months of the year the breeze is fresher there than in the London streets; the steamboat piers were conveniently situated; the fares were exceedingly low; the transit was comparatively rapid. There must have been tens of thousands of persons for whom river steamboats formed the most convenient, the cheapest and the most expeditious method of transit. Yet the

THE RIVER THAMES STEAMBOAT COMPANY (Limited).— Messrs. FULLER, HORSEY, SONS, and CASSELL are instructed by the Directors of the River Thames Steamboat Company (Limited) to offer for SALE by tender, as a going concern, in one lot, on Tuesday, June 1, 1886, the BUSINESS and the whole of the valuable PROPERTIES of the company, comprising:

1. Roff's Wharf, Woolwich, with pier to the landing stage, brick-built offices, manager's dwelling house, with wharf in front; dry dock, building slip, stores, smithy, and engineer's shop, fitted with valuable fixed plant and machinery.

2. Globe Wharf, Woolwich, adjoining, fronting the Thames, storehouses and dwelling house.

3. The Gridiron in front of the two preceding wharves, 284ft. long.

4. Extensive Premises at Battersea, with wharf, on the Thames, 330ft. long, dry dock, stores, and the other necessary buildings, and the valuable fixed plant and machinery.

5. The Piers and Dummies at Charlton, Cubitt Town, Putney, Battersea-square, Battersea-yard, Battersea Bridge, Battersea Park, Nine Elms; and the fittings and right of user at the following piers, viz.: Blackwall, Greenwich, West India Docks, Limehouse, Commercial Dock, Tunnel, Cherry-garden, London Bridge (Middlesex and Surrey), Kew, Wandsworth, Chelsea, Battersea Park Railway, Pimlico, Millbank, Lambeth, Westminster, Charing-cross, Waterloo, Temple, Blackfriars, Paul's Wharf, and All Hallows.

6. The Fleet of Vessels, consisting of 57 paddle-wheel steam vessels adapted for the passenger service of the Thames between London and Hampton Court on the up-river service and below bridge to Greenwich, Woolwich, Gravesend, Sheerness, Southend, Clacton-on-Sea, and Harwich. 39 of these vessels have, or shortly will have, Board of Trade certificates, and include the fast and favourite vessel, the Glen Rosa, on the Clacton service; the Alexandra, Albert Edward, Duke of Edinburgh, and Queen of the Orwell, on the Gravesend, Southend, and Sheerness line; and the Rosalind, Celia, and Orlando, built by Sir W. Armstrong, Mitchell, and Co., in 1855,* and fitted with all recent improvements for light draft navigation; and there are 18 vessels laid up in the docks and yards. In addition to the above fleet there are two small launches used for the business purposes of the company.

The average amount of gross receipts during the last four years has been £83,896 per annum, and there is little doubt that with the necessary increase of capital a very profitable and lucrative trade can be carried on. The above figures point to the transport by the company's vessels of many millions of passengers per annum, and it cannot for one instant be imagined that such a traffic should cease to exist, especially whilst it has the advantages of the various piers and landing-places, many of them belonging exclusively to the company, and so able to defy all competition. In addition to the river traffic there is a considerable revenue from the steam ferry boats between North Greenwich and Greenwich, and from the tolls paid by the Great Eastern Railway Company, whose boats carry passengers from North Woolwich to the Woolwich pier. A purchaser will be entitled to the benefit of the compensation clauses in the Metropolitan Board of Works Additional Powers Act, 1885, when a free ferry between North Woolwich and Woolwich is established by the board.

Printed particulars of the property, with the proposed arrangements for the inspection of same, and forms of tender will be forwarded on application either to the offices of the company, Adelaide-buildings, London Bridge, E.C.; Messrs. Stretton, Hilliard, and Co., 75 and 76, Cornhill, E.C.; or at the auctioneers' offices, 11, Billiter-square, E.C. The tenders will be opened at the offices of Messrs. Fuller, Horsey, Sons, and Cassell, Auctioneers and Surveyors, 11, Billiter-square, London, E.C., on Tuesday, June 1, at one o'clock precisely. The vendors do not bind themselves to accept the highest or any tender.

* This date should presumably be 1885 not 1855.

successive companies were the reverse of prosperous.

In June, 1886, the whole assets of the company were offered for sale by tender, and we reproduce the interesting advertisement that appeared in the London press, as it is a comprehensive record of the assets—and the liabilities—of the company.

This advertisement did not find a purchaser and two newspaper cuttings, unfortunately not dated, in the author's possession read, " A meeting of the Debenture-holders of the River Thames Steamboat Company (formerly the London Steamboat Company) is called for Thursday, to consider its position. It has had two of the finest seasons ever known, and nevertheless urgently wants more money. Some £67,000 of capital has been put into it since it took over from the old concern little more than a year ago, which was then bankrupt, and this is the result. It is a strange thing that the busiest river in the world cannot keep alive a passenger boat service without the loss of great sums every year "; and, " Yesterday afternoon only six steamers were running between London Bridge and Lambeth Pier, and all the employees have been given a week's notice to leave."

The foregoing two cuttings presumably appeared late in 1887, for a leading article in the *Standard* newspaper of November 9th, 1887, reads:

" The cessation of the regular service of passenger steam boats on the Thames between Battersea and London Bridge and down to Woolwich has caused no little inconvenience to a large number of the inhabitants of the Metropolis. We stated a few days ago that fourteen of the steam boats belonging to the Steam-boat Company had been sold to Mr. Moss Isaacs to be broken up for old iron. But the event is one which need excite no concern as to the future of the steam-boat traffic. On the contrary, it is a matter for congratulation that a number of old and crazy boats are thus finally disposed of, the concern having the benefit of as much money as the bare materials will fetch. The whole lot were worn out, and had been lying by for some time, forming no part of the active fleet. With regard to the present suspension of the boat service, it may be stated that for a long series of years the running of the boats between Battersea and Woolwich during the Winter months has been unremunerative, and it has been a matter of discussion at the Board of Directors of the Steam-boat Company whether it would not be advisable to suspend the entire service during the Winter season, with a view to lessen the loss. The undertaking has for some time past been subject to an order of liquidation. Last Winter a reduced service was in operation, but the experience gained in carrying out this arrangement was such that it was considered better, in the interests of all concerned, to suspend the service entirely during the Winter months. This plan is at present being adopted. The vessels now laid up, and that were running during the Summer season, require only the usual repairs in order that they may be made fit to recommence the service, which they could do at short notice. The number of boats engaged in the river service last Summer was thirty, several others being left aside owing to the abandonment of the long-distance service—as to Southend and Clacton-on-Sea—this part of the working being found unprofitable.

" A statement has gained currency that all the hands recently employed by the Thames Steam-boat Company have been discharged. This is an exaggeration. A large portion of the staff has been retained under an arrangement which gives partial employment. Neither are the boats entirely laid up. The ferry service between Greenwich and the Isle of Dogs in conjunction with the Great Eastern Railway is being maintained, as also the ferry between Port Victoria and Sheerness, under an arrangement with the South-Eastern Railway Company. Boats are likewise required for occasional service, as in the conveyance of emigrants from Blackwall to Gravesend. It is not unlikely that in the course of a week or two one of the steamers will be engaged in transferring troops from Shoeburyness to Woolwich. A few days ago one of the boats conveyed the members of the Metropolitan Board to Barking Creek, on a visit of inspection to the great precipitating works in course of construction at the northern outfall. The hope is entertained that the regular service will be revived in the Summer of 1888. How this is to be done is not yet clear, but it is scarcely to be credited that the Thames is to cease its functions as a regular Metropolitan highway. The steam-boat fleet comprises about forty vessels, including three of a superior description recently built for the Company by the firm of Armstrong and Co., at Newcastle. These three vessels show a distinct advance on their predecessors, and from the time they commenced running were great favourites with the public. They are built of steel, and have com-

P.S. " Celia " (1883) passing old Lambeth Bridge in 1887

P.S. " Jupiter " (1877) at sea, September, 1900

Face page 64

P.S "Princess of Wales" (1878), renamed "Bembridge"

P.S. "Glen Rosa" (1877)

pound engines with surface condensers, effecting a great saving of fuel. Unfortunately, the speed inherent in these three boats cannot be turned to full account, as they have to keep their place in the programme along with other and slower craft. But the accommodation on board is excellent, the saloon being of good height, with windows giving a full view of the river, and having a raised deck overhead. If the entire flotilla could be transformed into Armstrong boats the popularity of the service would doubtless be increased. Higher speed could be accomplished, and the comfort of the passengers promoted. In regard to the style of the boat, there are certain physical difficulties in the way of any great increase in size. The vessel's draught is limited by the frequent shallowness of the water. Breadth is affected by the bridges, and by the circumstance that the paddle-wheel is found to have certain advantages over the screw. Twin-screw propulsion would facilitate manœuvring (*sic*), but with a heavy deck-load of passengers, often congregating on one side of the boat, there would be some danger of capsizing. The tendency to heel-over beyond a certain angle is checked by the deeper immersion of the paddle-wheel on the lower side. In respect to the height which may be given to an upper deck a limit is at once fixed by some of the up-river bridges. It was even alleged at one time that the Armstrong boats were not safe for persons on the upper deck when passing under Hammersmith Bridge, and measurement was made to prove the contrary.

" Some of the causes which have contributed to embarrass the Thames Steam-boat Company are altogether external. Circumstances have altered immensely since the period when steam boats first began to ply on the Thames. In recent years certain changes have been particularly inimical to the passenger traffic on the river. One of the most potent forms of opposition has consisted in the extraordinary low omnibus fares arising out of the rivalry between the Road Car Company and the General Omnibus Company. The penny ride from Charing-cross to Liverpool-street has made many people forget the very existence of the steam boats. An equally low fare from Chancery-lane to Victoria also takes effect. The omnibuses have thus been put into severe competition both with the river boats and the District Railway. On the other hand, the steam-boat traffic at the Old Swan Pier has been interfered with by the completion of the Inner Circle Railway. For persons who are in haste the District Railway has the preference over the steam boats, apart from the question of price. All this is a new order of things compared with the days when omnibuses charged threepence between Charing-cross and Temple Bar, and no District Railway existed. At that date the penny steam-boat fare from London Bridge to Lambeth Pier commended the river route to the great mass of the people. The Steam-boat Company is at some disadvantage from the circumstance that while the road traffic has been relieved in the matter of its taxation, the pier charges remain undiminished. In addition to its own and other piers, the Company pays at least £6,000 a year for its use of the piers belonging to the Thames Conservancy. To meet this charge alone, independently of all other expenses, the Company must have the revenue arising from a million and a half penny fares. The Company also suffers from a change in the condition of the river. A trip down to Greenwich and Woolwich is discouraged by the ill-repute of the Thames in the vicinity of the main drainage outfalls. Up the river, from Kew to Teddington, there is a deficiency of water. This latter defect may be remedied in part, should the project of a lock at Isleworth be put into execution. So also it may be hoped that the new works which the Metropolitan Board are carrying out in connection with the outfalls, at a cost of some three-quarters of a million, will purify the Thames below Blackwall. As for the omnibuses and the railway, the only plan for the coming Steam-boat Company is to make the boats as attractive as possible, and, if it can be done, to increase the rate of speed. The open river is pleasanter than the dark and smoky tunnel, and the easy motion of the boat is more agreeable than the jolting of the 'bus, with its perpetual pulling up and going on again, to say nothing of the squeezing which prevails inside. The omnibuses have undergone great improvement in their construction, and the carriages on the District Railway are being more comfortably fitted. The people of London are gaining in the matter of locomotion; but they cannot afford to dispense altogether with the steam-boat service, accommodating—despite all opposition—about five millions of passengers in the year."

A new company was registered called the London River Steamers Ltd. with a capital of £250,000 in £5 shares, but we do not think it ever got further than registration.

The River Thames Steamboat Company continued in existence until, in 1890, the surviving boats were re-registered under the ownership of another company—The Victoria Steamboat Association. This company existed for a few years and went into liquidation and the sad tale can be gathered from the following list of altered registrations.

STEAMBOATS TAKEN OVER FROM WOOLWICH COMPANY.

ALBERT EDWARD	1876-1884	London Steamboat Company.
ALEXANDRA	1885-1888	River Thames Steamboat Company.
	1889	Names removed from the Mercantile Navy List.
ARIEL	1876-1884	London Steamboat Company.
	1885	River Thames Steamboat Company.
	1886	Name removed from the M.N. List.
CAMELIA	1876-1884	London Steamboat Company.
	1885-1889	River Thames Steamboat Company.
	1890-1894	Victoria Steamboat Association.
	1895	Name removed from the M.N. List.
CUPID	1876-1884	London Steamboat Company.
	1885-1889	River Thames Steamboat Company.
	1890-1896	Victoria Steamboat Association.
	1897	Victoria Steamboats, Limited.
	1898	Thames Steamboat Company.
	1899	Name removed from the M.N. List.
DORIS	1876-1884	London Steamboat Company.
	1885-1889	River Thames Steamboat Company.
	1890	Name removed from the M.N. List.
DUKE OF CAMBRIDGE	1876-1884	London Steamboat Company.
DUKE OF EDINBURGH	1885-1889	River Thames Steamboat Company.
	1890-1896	Victoria Steamboat Association.
	1897	Victoria Steamboats, Limited.
	1898	Names removed from the M.N. List.
DUKE OF CONNAUGHT	1876-1884	London Steamboat Company.
DUKE OF TECK	1885-1888	River Thames Steamboat Company.
	1889	Names removed from the M.N. List.
FAIRY	1876-1884	London Steamboat Company.
HERON	1885	River Thames Steamboat Company.
NYMPH	1886	Names removed from the M.N. List.
MARTIN	1876-1884	London Steamboat Company.
METIS	1885-1888	River Thames Steamboat Company.
PETREL	1889	Names removed from the M.N. List.
NORTH KENT	1876-1883	London Steamboat Company.
OREAD	1884	Names removed from the M.N. List.

NAIAD	1876-1884	London Steamboat Company.
	1885-1889	River Thames Steamboat Company.
	1890-1894	Victoria Steamboat Association.
	1895	Name removed from the M.N. List.
PLOVER	1876-1884	London Steamboat Company.
	1885-1889	River Thames Steamboat Company.
	1890	Name removed from the M.N. List.
PRINCE ALFRED	1876-1884	London Steamboat Company.
	1885-1889	River Thames Steamboat Company.
	1890	Victoria Steamboat Association.
	1891	Name removed from the M.N. List.
PRINCE ARTHUR	1876-1884	London Steamboat Company.
	1885-1889	River Thames Steamboat Company.
	1890-1893	Victoria Steamboat Association.
	1894	Name removed from the M.N. List.
PRINCE OF WALES	1876-1884	London Steamboat Company.
	1885-1889	River Thames Steamboat Company.
	1890-1896	Victoria Steamboat Association.
	1897	Victoria Steamboats, Limitied.
	1898-1908	Thames Steamboat Company.
	1909	Name removed from the M.N. List.
PRINCESS ALICE	1876-1878	London Steamboat Company.
	1878	Lost in collision in the Thames, 3/9/1878.
QUEEN OF THE ISLES	1876-1884	London Steamboat Company.
	1885	Sold to South Milton Iron Works, Gravesend.
QUEEN OF THE ORWELL	1876-1878	London Steamboat Company.
	1885-1890	River Thames Steamboat Company.
	1891-1896	Victoria Steamboat Association (renamed FAIRY QUEEN, 1891).
	1897	Victoria Steamboats, Limited.
	1898	Sold to H. A. Hill. London.
QUEEN OF THE THAMES / SEA SWALLOW	1876-1884	London Steamboat Company.
	1884-1888	River Thames Steamboat Company.
	1889	Names removed from the M.N. List.
SIBYL	1876-1878	London Steamboat Company.
	1879	Name removed from the M.N. List.
VESTA	1876-1884	London Steamboat Company.
	1885-1889	River Thames Steamboat Company.
	1890-1896	Victoria Steamboat Association.
	1897	Victoria Steamboats Limited.
	1898	Name removed from the M.N. List.

STEAMBOATS TAKEN OVER FROM W. H. STRATTON, LONDON, 1876.

BLACK PRINCE	1876-1884	London Steamboat Company.
	1885-1889	River Thames Steamboat Company.
	1890	Victoria Steamboat Association.
	1891	Name removed from the M.N. List.

GEORGE PEABODY	1876-1884	London Steamboat Company.
	1885	River Thames Steamboat Company.
	1886	Name removed from the M.N. List.
RIFLEMAN	1876-1884	London Steamboat Company.
	1885-1889	River Thames Steamboat Company.
	1890-1896	Victoria Steamboat Association.
	1897	Victoria Steamboats, Limited.
	1898-1908	Thames Steamboat Company (re-named NERISSA, 1899).
	1909	Name removed from the M.N. List.
RIVER QUEEN	1876-1884	London Steamboat Company.
	1885-1889	River Thames Steamboat Company.
	1890-1896	Victoria Steamboat Association.
	1897	Victoria Steamboats, Limited.
	1898-1908	Thames Steamboat Company.
	1909	Name removed from the M.N. List.
VICTORIA	1876-1884	London Steamboat Company.
	1885-1889	River Thames Steamboat Company.
	1890-1896	Victoria Steamboat Association.
	1897	Victoria Steamboats, Limited.
	1898-1899	Thames Steamboat Company.
	1900	Name removed from the M.N. List.
VOLUNTEER	1876-1884	London Steamboat Company.
	1885-1889	River Thames Steamboat Company.
	1890	Victoria Steamboat Association.
WARRIOR	1891	Name removed from the M.N. List.

STEAMBOAT TAKEN OVER FROM J. F. D'OYLY, LONDON.

ECLAIR	1876-1884	London Steamboat Company.
	1885-1888	River Thames Steamboat Company.
	1889	Name removed from the M.N. List.

PARTICULARS OF THE ABOVE BOATS.

Name	Built.	Where built.	Dimensions.	Gross tons.	N.H.P.
BLACK PRINCE	1862	Battersea	109.7 × 14.0 × 6.4	62	24
GEORGE PEABODY	1862	Nine Elms	109.7 × 14.0 × 6.4	62	24
RIFLEMAN	1861	Nine Elms	106.2 × 13.8 × 6.5	62	24
RIVER QUEEN	1856	Battersea	112.1 × 13.9 × 6.3	74	24
VICTORIA	1860	Nine Elms	114.9 × 13.9 × 6.5	96	20
VOLUNTEER	1861	Nine Elms	105.0 × 13.9 × 6.4	60	20
WARRIOR	1862	Nine Elms	107.5 × 13.9 × 6.3	60	24
ECLAIR	1865	Port Glasgow	179.8 × 20.2 × 8.5	237	120

A contemporary description of VICTORIA gives her as of "41 tons (net) register, fore and aft cabins, two oscillating steam engines with 22 in. cylinders, 24 in. stroke, and midship steering."

A handbill, dated August, 1872, advertises the boats running from London Bridge Wharf as follows: QUEENS OF ORWELL and THAMES to Ipswich; SEA SWALLOW, PETREL and OREAD to Southend; and ALBERT EDWARD and ALEXANDRA to Gravesend.

ECLAIR, 1865/1876.

This was a sea-going steamer built at Port Glasgow in 1865 and taken over from J. F. D'Oyly. She was by far the largest, being 237 tons gross. The engines were built by W. Simons & Co., Renfrew. There were two diagonal cylinders 40 in. by 42 in., and the N.H.P. was 120. For what purpose the ECLAIR was originally built, there appears to be no record.

LALLA ROOKH and WAVERLEY, 1846/1876.

In 1876, the London Steamboat Company acquired two steamboats named LALLA ROOKH and WAVERLEY, although neither of these boats appears in the Mercantile Navy List prior to 1875, except that in 1865 the LALLA ROOKH is shown as then owned by S. D. Sassoon, London. Both were built in 1846 at Blackwall, and were of the following dimensions:—

LALLA ROOKH	106.8	× 12.4	× 6.1;	gross tons 50;	N.H.P. 24;	Iron.
WAVERLEY	109	× 12.6	× 6.2;	gross tons 52;	N.H.P. 24;	Iron.

The former remained in service until 1879; the latter was disposed of in 1877 to Mr. G. A. Cape, London.

CHILDE HAROLD, 1862/1877.

The CHILDE HAROLD, built at Nine Elms in 1862, was purchased by the L.S.B. Co. in 1877. This steamboat passed through the hands of several owners. From 1871-1874 she was owned by L. H. Haslewood, London; in 1875 by W. H. Stratton and in 1876 by G. A. Cape. Her dimensions were: 106.5 ft. by 12.5 by 6.2; Gross tons 47, N.H.P. 24, built of iron. She remained in service until 1883.

By 1878, river traffic on the Thames had so much increased—prior to the September disaster—that five new boats were put on order. There were the first boats built for the London Steamboat Co., and were as follows:—

BRIDEGROOM and WEDDING RING, 1878.

The BRIDEGROOM and WEDDING RING were built to replace their namesakes of 1839 and 1844 respectively, and were engaged on short pier to pier trips. Both were built at Poplar, and were 111 ft. long by 13.8 by 5.8; Gross tons 77, and engines of 24 N.H.P. The BRIDEGROOM remained in service until 1900 and the WEDDING RING until 1890.

OSPREY and PRINCESS MARY, 1878.

The OSPREY and PRINCESS MARY were sister boats built for the longer downstream trips. The former was built at Cubitt Town and the latter at Poplar. They were much larger than the two up-river boats, being 131.3 ft. in length by 16 by 6.4, of 100 tons gross and 35 N.H.P. engines. The OSPREY remained in service until 1909 and the PRINCESS MARY until 1908.

PRINCESS OF WALES, 1878.

The PRINCESS OF WALES was the last and the largest of the five, and had the shortest period on the river. She was built at Woolwich by the London Steamboat Company, of the following dimensions: 137.4 ft. by 16.1 by 6.8; Gross tons 104, N.H.P. 40. She had oscillating engines, the two cylinders being 24.5 in. by 30 in., made by Messrs. Kincaid & Donald, Greenock. In 1886 she was sold to the Southsea & Ventnor Steamship Company and was renamed BEMBRIDGE, running between Southsea and the Isle of Wight until about 1907 when she was sold to a Portsmouth owner. After 1913, her name does not appear in the Register. She was re-boilered in 1894.

LILY and ROSE, 1880.

The LILY and ROSE, sister boats, both built at Deptford in 1880 continued in service on the river until 1908. In 1899, during the ownership of the Thames Steamboat Company, the LILY was renamed DIANA. Their dimensions were 110 ft. by 13.8 by 6.0; Gross tons 63, N.H.P. 24.

VALE OF CLWYD, 1865/1881.

In 1881, the Thames & Channel S.S. Company purchased the Clyde steamer VALE OF CLWYD, and in the following year sold her—with the GLEN ROSA—to the London Steamboat Company, who ran her on the Thames until 1888 under her original name. The VALE OF CLWYD was constructed of wood by T. B. Seath & Co., Rutherglen, in 1865, for the Dunoon and Ayr summer traffic. In the summer of 1877 she was chartered by Campbell and Gillies to run on the Wemyss Bay route, following on from Glasgow to Largs and Millport. The VALE OF CLWYD had machinery of a peculiar type, the engine being a " steeple " in combination with a small diagonal. Dimensions 186.5 ft. by 18.1 by 7. Gross tons 173. N.H.P. 88. Steam supplied by one haystack boiler.

GLEN ROSA, 1877/1883.

The GLEN ROSA was built by Caird & Co., Greenock, in 1877, for Messrs. Shearer and Ritchie, to run to Arran via Rothsay and Kilchattan Bay; but as it frequently happened with privately-owned Clyde steamers, the enterprise was a failure. So in 1881 the GLEN ROSA came to the Thames, having been purchased by the Thames & Channel Steamship Company, London. In 1888, on one occasion, she made the trip from London to Clacton and back on the same day, that being the first occasion on which an excursion steamer had done so. The trip was not remarkable for speed, but merely because the GLEN ROSA was scheduled to turn back at Clacton. She gained a good deal of publicity by attaining a speed of 21 knots with wind and tide in her favour. In her early days on the Thames, the author well remembers the common remark " GLEN ROSA late again "; this frequently happened, due to mechanical troubles.

Unlike most excursion steamers, the GLEN ROSA was not tied to any one route, but was run on various services. At one time, for example, she was put on between Harwich and Yarmouth, so as to connect with the KOH-I-NOOR at Harwich. She also ran from London to Southend and Sheerness.

Her owners—which were many—after coming to the Thames were as follows:—

1881 The Thames & Channel Steamship Company.
1883 The London Steamboat Company.
1885 The River Thames Steamboat Company
1890 The Victoria Steamboat Association.
1893 The Hastings, St. Leonards & Eastbourne Steamboat Company.
1896 A. Campbell, Bristol.
1898 P. & A. Campbell & Co., Bristol.
5/1917-5/1919 Taken over by the Admiralty and engaged as a mine-sweeper.
1921 Broken up by Pugsley & Co., Bristol.

The dimensions of the GLEN ROSA were as follows: Length 206.1 ft. by 20.1 by 7.5. Gross tons 254. Single-cylinder diagonal engine, the cylinder being 50 in. by 72 in.; N.H.P. 129. One haystack boiler, working pressure 50 lb. per sq. in. New engine 1887; new boiler 1891.

SHAMROCK and THISTLE. 1882.

These were the last boats to be built for the London Steamboat Company, both were built at Battersea and continued in service until 1908. Their dimensions were: 113 ft. by 14 ft. by 6.5 ft. Gross tons 71. N.H.P. 24.

ORLANDO and ROSALIND. 1885.

These two sister boats were built at Low Walker on the Tyne and used entirely on pier to pier trips. They were the first boats built to the order of the River Thames Steamboat Company and differed from earlier " penny-boats " in having large well-lighted saloons aft and forward, and the engine controlled from the deck under the bridge. The engine was two-stage expansion with inclined cylinders 15 in. and 30 in. diameter by 24 in. stroke, steam at 100 lb. per square inch pressure was supplied by a cylindrical return tube boiler with two furnaces. When driving the paddle wheels at 44 revs. per minute, the I.H.P. was 100, the speed 10 knots and the coal consumption about 1.2 tons in twelve hours. The wheels were 9 feet in diameter and each had seven curved feathering floats 4.6 ft. by 1.8 ft. deep. The loaded draught was 3.5 ft. The crew consisted of a master, a mate, two deckhands, an engineer and a stoker. They were certified to carry 300 passengers. The Orlando continued in service until 1908, but the Rosalind was withdrawn in 1905.

Cardinal Wolsey. 1889.

This was the largest steamer ever built for up river or " above bridge " service, and the first to be constructed of steel for such work. Constructed by Edwards & Symes, London, she had two oscillating cylinders 20 in. diameter by 21 in. stroke. This boat remained in service on the Thames until 1909.

The Shah, Kaiser and H. M. Stanley. 1889.

These three were built at Poplar and the last constructed for the River Thames Steamboat Company. The funnels of this company were painted white with two red bands at the top. The Shah was renamed Penelope in 1899 and continued in service until 1910. H. M. Stanley was renamed Portia in 1899, and with Kaiser was withdrawn in 1909.

Lord of the Isles/Jupiter. 1877/1890.

On the Clyde, the year 1877 is memorable in the establishment by the Glasgow and Inveraray Steamboat

Company of a daily service to Inveraray and back by a new steamer LORD OF THE ISLES.

This vessel was launched on 30th May, 1877, from the Partick yard of D. & W. Henderson & Co. Built of iron, it was in dimensions 246 ft. by 24.2 by 9 ft.; gross tonnage 451; and was fitted with diagonal oscillating engines having two cylinders 46 in. diameter by 66 in. stroke; n.h.p. 265.

She was one of the most beautiful paddle steamers ever built and was the first excursion steamer to be fitted with electric light. In 1890 she was purchased by the Victoria Steamboat Association and attracted a great deal of attention on her daily trips down the Thames. In fact, it may be claimed that the big improvement in the standard of the London river steamers dates from her arrival.

In 1896, she was transferred to Mrs. C. Black (Mr. B. P. Black, Manager) and renamed JUPITER. She ran from London to Southend and Margate for several seasons. After being laid up for some time she returned to the Clyde in 1904, having been purchased by the Isles Steamship Company, who changed her name once again, this time to LADY OF THE ISLES. She did very little work after her return to her native waters and was cut up at Dumbarton in 1905.

Mr. L. G. Lane, in his interesting little publication "Down the River to the Sea," 1934, says of the LORD OF THE ISLES "In spite of her romantic name this vessel became notorious to the public through her numerous breakdowns. Day excursions on this boat were often night excursions as well, as passengers frequently returned home in the early hours of the next morning. But her greatest achievement was when she collided with London Bridge and smashed both her funnels. Apparently, she had seen the Palace Steamers go under the bridge with impunity and thought she could do the same."

Bismarck and Empress Frederick. 1891.

Following the purchase of Lord of the Isles, the Victoria Steamboat Association had two new boats built at Poplar similar to Kaiser; they were named Bismarck and Empress Frederick. The former was renamed Phyllis in 1899 and continued on the river until 1909; the was withdrawn in 1911.

The Mermaid. 1891.

In 1891, Samuda Bros., Poplar, built for the Victoria Steamboat Association the paddle steamer The Mermaid. Dimensions: Length 160 ft. by 20.1 by 7.9. Gross tons 194. She had two oscillating cylinders 30 in. diameter by 33 in. stroke. N.H.P. 52. Between the date of her building and the time she was cut up in 1927 this paddler had many changes of ownership. In 1897 she was transferred to the Thames Steamboat Company and remained with them until 1912 when she was acquired by the Rosherville Pier and Steam Packet Co. In 1916 The Mermaid was purchased by the Goole and Hull Steam Packet Co. and in the same year was chartered for temporary service by the North Eastern Railway for use on the Humber Ferry Service. After the first world war she made trips during 1919 and 1920 on the Humber being then owned by the Alfred Calvert Shipping Co. of Goole.

In 1922, a Mr. Elton of Limerick took over The Mermaid and ran her between Limerick and Kilrush, a 45 mile run down the river Shannon, but after a few trips she was commandeered by the Irish Free State Army. After release the vessel was laid up for a considerable time until she left for Belfast to run trips on Belfast Lough to Bangor, but this project was unsuccessful and on the 25th November, 1927, The Mermaid left Belfast for Preston for scrapping by Messrs. Ward the shipbreakers.

Orchid and Palm. 1892.

These two boats were built by R. & H. Green, of London, for "local" or pier-to-pier work. The engines had two oscillating cylinders 21.5 in. diameter and 22 in. stroke (N.H.P. 24) and were by J. Penn & Son of Greenwich. Palm ran until 1912 and Orchid until 1913.

Alexandra, Boadicea and Cleopatria. 1898.

In 1898 a new concern, the Thames Steamboat Company, took over the old Victoria Association fleet, and they immediately ordered three new boats. They were the only new boats this company ever ordered and were the last new boats put on " above bridge " services for many years. They were built of steel, had a loaded draught of 3 ft. Tonnage: gross 111, net 70. Main deck flush with saloons below. Speed 10 knots. Fuel consumption 8 cwt. per hour. Bunker capacity 10 tons. Two oscillating cylinders 23 in. by 24 in. working at 50 lb. pressure. These three boats were built and engined by the Thames Iron

THE THAMES & CHANNEL STEAMSHIP CO., LTD.

2, Adelaide Buildings, London Bridge

STEAM BOAT ARRANGEMENTS

Clacton-on-Sea, Walton-on-the-Naze, Harwich, Ipswich.

By the new and magnificently-fitted Steamer "GLEN ROSA" (speed 22 miles per hour); "VALE OF CLWYD," "QUEEN OF THE THAMES," "QUEEN OF THE ORWELL," or "DUKE OF EDINBURGH," from London Bridge (Old Swan Pier). See daily papers or handbills.

Rosherville Gardens, Gravesend, Southend-on-Sea, Sheerness.

By the magnificently-fitted Saloon Steamers "ALEXANDRA," "ALBERT EDWARD," "DUKE OF CONNAUGHT," DUKE OF CAMBRIDGE," "SEA SWALLOW," "PRINCESS OF WALES," "PETREL," "METIS," ETC., ETC., from London Bridge (Old Swan Pier). See daily papers and handbills.

SEA TRIPS

By the magnificent sea-going Steamer "ECLAIR" from Blackwall calling at NorthWoolwich and Gravesend. Steamers from London Bridge (Old Swan Pier) to Blackwall in connection with this steamer.

For particulars see daily papers or handbills.

Calais, Boulogne, Dunkirk, Deal and Dover.

By the new and magnificent sea-going Steamer "GLEN ROSA" or "VALE OF CLWYD" from Ramsgate. For further particulars see handbills.

Woolwich and Greenwich.

Daily, by Steamers every half-hour, from Westminster, Charing Cross, Temple, Blackfriars, and London Bridge (Old Swan Pier). Extra Steamers and Express Service of Steamers. See handbills.

Kew Gardens, Wandsworth, Putney, and Hammersmith. Every half-hour, from Chelsea (Cadogan Pier). Steamers from London Bridge and all intermediate Piers to and from Chelsea every 10 minutes.

Hampton Court, Richmond, Kingston or Teddington.

By the magnificently-fitted Saloon Steamer, "CITIZEN-S," specially built for the shallow water up the river, from London Bridge (Old Swan Pier), calling at all piers up the river. See handbills or daily papers.

Special or Private Steamers.

May be engaged for River Excursions. Apply to the offices of the Company for Tariff or further particulars.

All kinds of refreshments provided on board.

July, 1881. Edgar Shand, Manager.

A contemporary Poster.

Face page 80

P.S. " Thistle " (1882) approaching Albert Bridge in 1887

P.S. " The Shah " (1889) at Lambeth Pier

Works, Blackwall. They continued in service until 1912 when the company ceased operating. Subsequently CLEOPATRA ran for a time on Belfast Lough, later on the North East Coast and finally on the Tay at Dundee.

STOUR. 1878/1900.

This was the last boat acquired by the Thames Steamboat Company. Originally owned by the Great Eastern Railway and employed on their Orwell river service, she was purchased by the Thames company in 1900 and worked on the Thames from 1901 till 1908. She had compound diagonal engines by Young & Son, London, with cylinders 26 in. and 60 in. by 45 in.

Name	Date	Builders	Dimensions (ft.)	Gross tons	N.H.P.
CELIA	1885	Armstrong Whitworth	100 x14 x6.1	71	30
ORLANDO	1885	do.	do.	71	30
ROSALIND	1885	do.	do.	71	30
PRINCESS BEATRICE	1888	Jarrow	115 x16.1x6.3	89	45
CARDINAL WOLSEY	1889	Edwards, Symes	150 x11.5x6.7	130	31
H. M. STANLEY renamed PORTIA	1889	Samuda Bros.	120.2x16.1x6.8	97	24
KAISER	1889	do.	do.	97	38
THE SHAH renamed PENELOPE	1889	do.	do.	97	38
BISMARCK renamed PHYLLIS	1891	do.	do.	97	24
EMPRESS FREDERICK	1891	do.	do.	97	24
ORCHID	1892	R. & H. Green	125 x20 x6.9	137	24
PALM	1892	do.	do.	137	24
ALEXANDRA	1898	Thames Iron Works	120.1x17.1x7.2	111	31
BOADICEA	1898	do.	do.	111	31
CLEOPATRA	1898	do	do.	111	31
STOUR	1878	do.	125 x17.6x6.8	112	30

CHAPTER IV

THE GENERAL STEAM NAVIGATION COMPANY.

THE G.S.N. Co., the pioneer major steamship company, was officially formed in October, 1824. The origin of the Company was due to a few merchants who owned a number of small steam packets trading between London and Margate and who arranged to combine for the purpose of developing steam navigation. One of the principal promoters was Mr. Thomas Brocklebank, a wealthy timber merchant, who had built in his own yard at Deptford a steam packet named the EAGLE which he ran between London and Margate.

At a meeting held in London on 11th June, 1824, it was agreed to form a Company to be designated The General Steam Navigation Company. The original object of this Company was to trade all over the World; actually, however, they limited their scope to the Home Trade (as it is still called), that is, the ports lying between Hamburg and Brest. At a later date the Company included in their activities the Mediterranean.

The Clyde Shipping Company claims to be the oldest steamship company in the World, but their vessels were built for river voyages alone, whereas the first vessels of the G.S.N. Co. were sea-going, and the latter Company can therefore claim to be the oldest existing sea-going steamship Company. In this record of Thames steamboats, the services dealt with are limited to those boats plying between London, Yarmouth, Thanet Resorts, and excursions to the Channel Ports.

The nucleus of the G.S.N. Company's fleet of Thames and estuary steamboats acquired in 1824 was comprised of the EAGLE and the ECLIPSE, both built in 1820, the VENUS and the HERO built in 1821, the ROYAL SOVEREIGN built in 1822 and the CITY OF LONDON and the BROCKLEBANK built in 1824. These seven packets had been built for and were engaged on the London-Thanet service. They were supplemented by the acquisition of the ATTWOOD in 1825, and soon afterwards by the COLUMBINE, the MAGNET and the HARLEQUIN in 1826, and the RAMONA in 1828. These twelve packets sufficed for some years, unless perhaps any of the sea-going vessels temporarily replaced them.

The Thames steamers of the G.S.N. Company may be divided into three periods, dating from 1824, 1848 and 1887.

An early General Steam Navigation Company's advertisement of a daily service to Herne Bay and Margate by the ECLIPSE stated that it would leave " Every weekday " from the Hungerford Market Pier at 7.30 a.m., calling " at Old Swan Pier at 7.45 a.m., Tunnel Pier at 8.0 a.m. " and Greenwich at 8.30 a.m. for Herne Bay and " Margate." It was announced that " directly the steamer " starts an excellent plain breakfast can be obtained in the " chief cabin, 1s. 0d. each. Ham, eggs, etc., are plentifully " supplied at a trifling additional charge. An excellent

G. S. N. CO.

Vessels Acquired 1824 - 1828

Name	Built	Acquired	Builder	Gross Tons	Mat.	Dimensions Ft.	Engines by	N.H.P
Eagle	1820	1824	Brocklebank, Deptford	170	Wood		Boulton & Watt	40
Venus*	1821	1824	Brent, Rotherhithe	202	do.	129.5 × 21.1	do.	70
Royal Sovereign	1822	1824	Brocklebank, Deptford	220	do.	135.3 × 19.2	do.	100
City of London	1824	1824	Brocklebank, Deptford	185	do.	126 × 19	do.	30
Hero	1821	1824	Brocklebank, Deptford	233	do.	130.5 × 20.1	do.	100
Brocklebank	1824	1824	Brocklebank, Deptford	125	do.	108 × 20		20
Eclipse	1820	1824	Brent, Rotherhithe	88	do.	104 × 16.9	Boulton & Watt	70
Attwood	1825	1825	Blackwall	144	do.	147 × 19.8		100
Columbine	1826	1826	Evenden, Deptford	241	do.	157.6 × 24.5	Butterley Co.	140
Magnet	1826	1826	Tebbutt, Limehouse	166	do.	139.1 × 21.2	Boulton & Watt	150
Harlequin	1825	1826	Evenden, Deptford	315	do.	138.1 × 22.4	Butterley Co.	140
Ramona	1828	1828	Evans, Rotherhithe	178	do.	148.8 × 22	Butterley Co.	100

* Wrecked 22nd April, 1831, by collision with H.M.S. Firefly. 280 passengers on board.

"band of music is on board." The ECLIPSE was due at Herne Bay at 12.45 p.m. and at Margate at 1.30 p.m. The return trip was scheduled to leave Margate at 1.45 p.m. and Herne Bay at 2.45 p.m., being due at Hungerford Market at about 8.0 p.m. The average speed was 12.5 knots.

Smiles, in his *Life of John Rennie*, says that Rennie in his efforts to convince the Admiralty as to the value of steam power "hired the Margate steamboat ECLIPSE to tow the HASTINGS from Woolwich to two miles beyond Gravesend against a rising tide. The experiment was made on 4th June 1819, and proved so successful that the Admiralty were induced to authorise a steamboat to be built for similar service." This date does not agree with the date given on page 83, but there may have been an earlier steamboat named ECLIPSE.

LITTLE WESTERN.

The LITTLE WESTERN was for many years a favourite vessel on the London, Margate and Ramsgate station. Her first appearance in Ramsgate Harbour in 1844 created quite a sensation and she was visited by many anxious to inspect the luxurious and sumptuous fittings of her cabins and saloon. She carried a gilded serpent at her bows. The LITTLE WESTERN finished her days as a coal hulk below Gravesend until she was finally cut down by a screw steamer coming up the Thames.

MAGICIAN.

The MAGICIAN, a continental steamer, was occasionally

employed between 1852 and 1856 on the Kent Coast resorts service.

ALBION. 1848.

The ALBION was the first of the second period excursion boats, being built at West Ham in 1848. The engines were constructed by the G.S.N. Company at their works at Deptford Creek.

PRINCE OF WALES. 1843/1849.

In 1849, the G.S.N. Company purchased from the Margate Steam Packet Company the paddle steamboat PRINCE OF WALES. After a complete re-conditioning which included new engines, boilers and feathering wheels she remained a popular favourite on the Margate and Ramsgate station for upwards of thirty years.

The PRINCE OF WALES had a clipper bow and was built at Blackwall in 1843. She originally had a beam engine that had previously seen service in an earlier Margate steamboat named the ROYAL GEORGE. At the time of her launching, she narrowly escaped disaster: as she was sliding down the ways, the cleat of the bow gave way, due to some of the bolts breaking, with the result that she failed to take the water properly. Eventually with the aid of screw-jacks and tugs, she was floated off.

EAGLE (2). 1853/1856.

In 1856, the Company purchased that very famous steamer, the EAGLE, a vessel whose name perpetuated the name of the Company's first vessel and which was the forerunner of other equally famous " Eagle " steamers. She was licensed for 466 passengers to Dover. She had four furnaces and her engines worked at a steam pressure of 25 lb. Her draft was 7 feet. Paddle wheels 16 ft. diameter, fitted with 12 feathering floats, gave her a speed of 14¼ knots.

The EAGLE remained on the service for some thirty years and was, indeed a real picture; her funnel painted all black, her lifeboats a deep blue, white paddleboxes and red wheels, while the figurehead of an eagle and name surrounded by gold filigree work adorned the bow. She had no main deck saloon, being open from stem to stern. Prior to about 1887 the Thanet boats made only a single trip, running down to Ramsgate one day, staying there for the night, and returning to London on the day following.

SIR WALTER RALEIGH. 1859/1862.

Built at Renfrew in 1858 and purchased by the Company in 1862, the SIR WALTER RALEIGH was used entirely between London, Margate and Ramsgate. She was broken up in 1891.

HILDA. 1862/1868.

In 1868, a steamboat named the EUGENIE was acquired. In her early days this vessel had a varied career. Built for the South Eastern Railway's Continental service between Folkestone and Boulogne she was found to be unsuitable for that class of traffic, only having been in service from 3rd May until 26th November, 1864. After being withdrawn a purchaser was soon found, and she crossed the Atlantic, finding service as a blockade runner. Eventually she returned to England and was purchased for the Thanet service, being re-named HILDA.

The HILDA was one of the most handsome of the Navigation boats, and with her clipper bow resembled the EAGLE, except that she had two funnels, one forward and one aft the paddleboxes. She was the only two-funnelled excursion steamboat that the Company ever had, apart from the ROYAL SOVEREIGN, purchased in 1929, and the ROYAL DAFFODIL built in 1939. She was re-boilered in 1869 and cut up in 1889.

HOBOKEN. 1873.

The HOBOKEN was put in service for the 1873 Thanet season. So far as the G.S.N. Company was concerned, she was an innovation, being the Company's first deck saloon steamer. She had saloons fore and aft the paddle-boxes, which connected, making a continuous promenade.

The HOBOKEN ran in conjunction with the EAGLE and the HILDA. She was broken up in 1897.

In the early days, for a time, the steamers after disembarking the majority of their passengers at Margate and Ramgate would proceed on to Deal and return to Ramsgate for the night. It was not usual for the three boats to run every day except at weekends when the traffic was heavier, one of them usually being known as the " husbands' boat."

CONCORDIA, SEINE, MOSELLE.

G.S.N. Co. Continental steamers were occasionally used in the 1870's on the Margate and Ramsgate service.

Face page 88

P.S. " Hoboken " (1873)

P.S. " Halcyon " (1889) entering Boulogne

P.S. " Eagle " (1898)

G. S. N. CO.

Vessels Acquired 1848 - 1873

Name	Built	Acquired	Builder	Dimensions (ft.)	Gross Tons
Albion	1848	1848	West Ham	174 × 23.2 × 13.2	338
Prince of Wales	1843	1849	Miller, Ravenhill, Blackwall	180 × 22 × 10	429
Eagle (2)	1853	1856	Northfleet	200 × 24.1 × 11.3	325
Sir Walter Raleigh	1858	1862	Renfrew	168.5 × 20.6 × 10.4	239
Hilda	1862	1868	Samuelson, Hull	225 × 24.2 × 11.9	428
Hoboken	1873	1873	R. Napier & Sons, Glasgow	222.4 × 22.9 × 10.5	413

Name	Engines by	Type	Cylinders in.	N.H.P.
Albion	G.S.N. Co., Deptford	Side Lever	48 × 42	125
Prince of Wales	From Royal George *	do.	46 × 42	136
Eagle (2)	G.S.N. Co., Deptford	do.	49 × 48	130
Sir Walter Raleigh	MacNab, Greenock †	Oscillating	38 × 42	100
Hilda	Samuelson, Hull	do.	58 × 54	220
Hoboken	Napier & Sons, Glasgow	Diag. Compound	40, 60 × 72	200

* New engines by G.S.N. Co. Cylinders 48 in. × 42 in. New boilers 1872.
† New boilers 1866.

HALCYON, MAVIS, ORIOLE, LAVEROCK, PHILOMEL. 1887-1889.

Fourteen years passed without the Company adding any paddle steamboats to its fleet, but when it was rumoured that other interests were considering building new faster rival steamers the G.S.N. Co. thought it was about time they built something more modern and between 1887 and 1889 there appeared the HALCYON, MAVIS, ORIOLE, LAVEROCK and PHILOMEL. These five steamers were of 17 knots, had compound diagonal engines, and were employed on the Kent Coast services and a new service to Great Yarmouth. All five were built by Messrs. J. Scott & Company, Kinghorn. The first to be launched was the HALCYON, and she ran her trial trip on the Forth on 9th July, 1887. All had saloon decks extending the full width of the vessel and carried for some distance forward of the paddleboxes. Originally they had two masts, but later the main mast was removed, which alteration did not add to their appearance. They had one funnel forward of the bridge. After the rival "Belle" steamers went into service, the colour of the G.S.N. excursion steamboats was changed from the standard black to cream or buff. Later a narrow black band was added.

Name	Built	Dimensions	G. Tons	N.H.P.	Cyls. ins.	Re-boilered
HALCYON	1887	209.2×26.2×9.1	458	240	33.5, 61×60	1899
MAVIS	1888	210.3×26.3×9.4	474	240	33.5, 61×60	—
ORIOLE	1888	230 ×27.6×9.8	484	260	34, 62×66	1899
LAVEROCK	1889	210.2×26.2×9.4	470	240	33.5, 61×60	1901
PHILOMEL	1889	236 ×27.1×9.5	564	327	37, 69×66	1900

These boats did good service for about twenty years and were ultimately disposed of as follows:

HALCYON: Sold 1904 to the South of England Steamboat Company. Re-purchased in 1906 by the G.S.N. Company and sold to the Hamburg-Stade-Atlander-Linie. Re-named CUXHAVEN and used as a river boat on the Elbe.

PHILOMEL: Sold 1907 to the Furness Railway Company and worked between Barrow and Fleetwood. Sold 1913 to Messrs. T. W. Ward & Co. for scrapping.

LAVEROCK: Sold 1908 to the Compagnie Maritime Bordeaux-Ocean of Bordeaux. Re-named VILLE DE ROYAN. Sold and dismantled 1922.

MAVIS: Last trip to Margate on 31st July, 1909. Sold 1910 to Pockett's Bristol Channel Steam Packet Company. Broken up 1915 at Briton Ferry.

ORIOLE: Sold 1912 to Holland and converted into a hulk.

EAGLE (3). 1898.

The G.S.N. Company had for many years practically a monopoly of the pleasure boat traffic between London and the Kentish Coast Resorts; their boats were fast and commodious, and the captains in command of them were all men of experience. In the closing years of the nineteenth century, however, several competitors appeared to challenge the G.S.N. Company's monopoly: so it was decided to build a larger and faster vessel which perpetuated the well-known name EAGLE.

The new steamer was a handsome 18-knot saloon vessel, built by Gourlay Brothers at Dundee in 1898. She commenced running in that season and continued to do so until she made her last trip to Margate on 26th August, 1928. On 5th July, 1909, she inaugurated a service from Tilbury calling at Southend, Margate, Ramsgate, Deal and Dover.

Like her immediate predecessors, the EAGLE originally carried two masts, but the main mast was afterwards removed. In 1929, she was sold to a firm of shipbreakers in Holland and subsequently her hull was converted into a landing-stage on the river Maas, above Dordrecht.

During the 1914-1918 War, the EAGLE was taken over by the Admiralty for use as a mine-sweeper and re-named AIGLON, commencing her war service on 22nd November, 1915, she remained under Admiralty control until 23rd July, 1920.

KINGFISHER. 1906.

Competition to Thanet and the East Coast resorts continued between the New Palace Steamers (Chapter V), the General Steam and the Belle Steamers (Chapter VII) with the periodical " butting in " of an outsider. All the Thanet trips were based on a daily return run, but the trip to Yarmouth was a day's run each way.

However, in spite of the many advantages of the paddle steamer over the screw for river traffic, the G.S.N. Company decided to have built a screw steamer for their cross-Channel work. Accordingly an order was placed with W. Denny and Brothers, Dumbarton, for a triple screw turbine-driven steamer.

This new venture of 1906, named the KINGFISHER, was beyond dispute a very fine production, as might be expected, coming from such famous builders of cross-Channel steamers. The KINGFISHER was the fastest vessel ever built for excursion traffic on the Thames, her speed being 21.5 knots, but nevertheless she was not a success on the London river. Her run was from Tilbury to Dover and Boulogne, calling at Southend, Margate, Ramsgate and Deal. Her first trip to Dover was on the 16th June, 1907, and she continued in service until 1911.

Two unforeseen factors made it advisable to withdraw the KINGFISHER while she was comparatively a new boat. The relatively high speed at which she travelled and the consequent wash created brought the Company a troublesome harvest of claims for damage from owners of small craft on the river. In addition, she proved difficult to manœuvre at the piers. As the pleasure piers are usually built on piles, they do not act as breakwaters. This means that any currents created by the tide have free play around the landing stage. It was found that the KINGFISHER, being a screw steamer, took much longer to come alongside the piers than her paddle-driven contemporaries.

These delays added considerably to the vessel's journey time, and together with the claims already mentioned influenced the Company in their decision to dispose of her after running her for only five seasons. The failure of the KINGFISHER served to demonstrate the supremacy of the paddle-wheel as a means of propulsion for vessels running on the Thames excursion services. Paddle-steamers are ideal for service on routes where a great deal of manœuvring has to be done very quickly, because the paddle wheels get a grip on the water much sooner than a screw propeller.

The KINGFISHER had two decks, the promenade deck extending from the bow to within a few feet of the stern; she had one funnel, and two masts. Her dimensions were: length 275 feet, beam 32.1, depth 10.2, gross tons 982; three steam turbines, 399 n.h.p., speed 21.43 knots.

In 1912, she was sold to owners at Trieste and took up duties in the Adriatic, being re-named VENEZIA. In 1913, she was fitted for oil fuel burning; later she was sold to the China New Era Shipping Company, Hong Kong.

GOLDEN EAGLE. 1909.

In the summer of 1909, Messrs. J. Brown & Co., Clydebank, launched the GOLDEN EAGLE, a vessel that was to see a great deal of active service during the 1914-1918 War. This new addition was a big advance on all her predecessors, not only in dimensions and tonnage, but also on account of the introduction of triple-expansion engines. She was one of the most popular boats running from London, doing the return trip daily during the summer months to Margate and Ramsgate, with occasional trips to Boulogne. In 1934, the GOLDEN EAGLE was fitted for burning oil fuel, and at the same time the cowl on the top of the funnel was removed.

In February, 1915, she was taken over by the Admiralty and converted into a troop carrying vessel. She left Deptford on the 18th February for Southampton, the same day on which the German Government declared a blockade of Great Britain. On 23rd February, she left Southampton for Rouen with 698 officers and men; and thenceforward plied continuously between Southampton, Havre, Rouen, Calais, Boulogne, Dunkirk, Folkestone, Cowes and Felixstowe. This famous paddle steamer had carried no less than 518,101 troops and other units across the Channel when she finally completed her service as a transport in November, 1919. A special brass plate to commemorate her war service was later affixed to the chart-room by order of the Admiralty.

GOLDEN EAGLE also served throughout the second World War during which she put into Weymouth for a major refit and holds the record as being the widest vessel to pass through the bridge to the inner harbour there. On being restored to the Kent Coast passenger services in 1947, she was fitted with an additional mast (a small stump main-mast). She principally runs on the London, Southend and Ramsgate direct service.

CRESTED EAGLE. 1925.

After 1925, the Company had the excursion business down the Thames Estuary entirely in their hands—so far as luxury steamers were concerned—for some few years, their old rivals having disappeared; nevertheless, the Company decided to build a still larger and more commodious paddle steamer. On 25th March, 1925, Messrs. J. Samuel White & Co. launched, at Cowes, the CRESTED EAGLE, a vessel 299 feet in length and over 1,100 tons gross, compared with the GOLDEN EAGLE's length of 275 feet and 790 tons gross. Like the GOLDEN EAGLE, the commodious promenade deck of this steamer ran from the bow to within a few feet of the stern post; the hull being divided into ten compartments by watertight bulkheads. The foremast was hinged and the funnel made telescopic for the purpose of passing under London Bridge to take up her station at the Old Swan Pier. A bow rudder assisted the steering when going astern. As in the GOLDEN EAGLE, the motive power was obtained from a fine set of triple-expansion engines, which could give a speed of over 18 knots.

The CRESTED EAGLE was the first Thames pleasure steamer to be fitted for burning oil fuel—in fact she was the first in Europe. Until 1931, she ran on the Ramsgate service, but in the 1932 season was transferred to the Clacton and Felixstowe run. In 1935, for the Silver Jubilee, her hull was painted a light biscuit colour, with green boot-topping. This colour scheme also applied to the GOLDEN and ROYAL EAGLES, but was not repeated in the following seasons.

In May, 1937, the CRESTED EAGLE took up her moorings in the Thames off the Embankment as a floating grandstand for viewing the Coronation procession of King George VI, on the 12th May.

In 1939, the CRESTED EAGLE was given a new funnel and was fitted with deck windscreens painted white.

It is to be regretted that this once popular steamer is no more, as in June, 1940, she was bombed, and sank while taking part in the famous Dunkirk evacuation. The

P.S. " Crested Eagle " (1925) with funnel lowered

T.S.M.V. " Crested Eagle " (1938)

P.S. "Royal Sovereign" (1893) leaving Margate on her last trip, 1929

P.S. " Royal Eagle " (1932)

vessel received a direct hit and rapidly became a blazing furnace, due to the ignition of the oil fuel in the bunkers.

ROYAL SOVEREIGN.

Purchased in March, 1929, from New Palace Steamers Ltd., she ran to Margate and Ramsgate for the 1929 season only. (See The Palace Steamers, Chapter V.)

ROYAL EAGLE.

On 24th February, 1932, the ROYAL EAGLE was launched from the yard of Messrs. Cammell Laird & Company, Birkenhead. This new vessel was built for the Margate-Ramsgate service, and although larger steamers had been on the Thames before none had been more popular and luxuriously fitted. The dining saloons will seat 310 persons, and the vessel can carry a total of 2,000. Although 10 feet shorter than the CRESTED EAGLE, her gross tonnage shews an increase of 400.

The ROYAL EAGLE has several entirely new features. She is the first Thames excursion steamer on which oil is used throughout for cooking, etc. She has a streamlined rudder aft and a bow rudder forward, as it is the practice of these boats to turn at a favourable opportunity on the way up the river at night, finishing the trip stern first. An important feature is the glass-enclosed deck saloon with a spacious sun-deck above, there being four decks altogether.

Before the Company finally decided to build a vessel propelled by steam and paddles every other form of power and propulsion was thoroughly examined. Although, generally speaking, the screw has superseded the paddle, the advantages of the latter lay in the much larger deck space available for the free movement of passengers, and also in the greater ease of navigation and manœuvring in a river crowded with moving barges.

The ROYAL EAGLE, unlike other EAGLES, has two masts, which considerably enhance her appearance. The funnel was originally all cream colour, but in 1938 had a black top added similar to the other EAGLES. The hull was painted white down to the level of the sponsons in 1939. She made her maiden trip on 14th May, 1932. It may be of interest to record that when the ROYAL EAGLE was launched a bottle of whisky was used and it is believed to have been the first occasion on which a vessel was named with whisky.

ROYAL EAGLE attended the Naval Reviews at Spithead in 1935 and 1937 and the maiden voyage departure of the Cunard White Star liner QUEEN MARY on the 27th May, 1936, on all three occasions operating from Southampton, and whilst there also made a cruise round the Isle of Wight. In 1946 the ROYAL EAGLE resumed the Kent Coast passenger service and in that year and the following season terminated at Margate.

A feature common to all the EAGLES was that the bridge was situated abaft the funnel, although an unusual practice these days this was done to assist stern-first navigation in the Thames above Greenwich.

When the second World War broke out, the ROYAL EAGLE became H.M. anti-aircraft vessel ROYAL EAGLE. In May, 1940, this famous vessel took part in the Dunkirk evacuation, and in the following November she shot down a dive-bomber during an attack on shipping in the Thames Estuary.

ROYAL EAGLE'S WAR RECORD

The following particulars are taken from *The Daily Telegraph* dated 1st August, 1943:—

Nights at sea - - -	520
Days at sea - - -	132
Mileage - - - -	25,000
Actions against aircraft -	52
Planes destroyed - - -	2
Lives saved - - -	24

The ROYAL EAGLE made three trips to Dunkirk, was dive-bombed 43 times and brought home 3,000 soldiers.

PARTICULARS OF "EAGLE" STEAMERS

Name	Built	Dimensions	G. Tons	Cyls. (in.)	H.P.
EAGLE (3)	1898	265 ×30.1×10.1	647	37, 70×60	N. 316
GOLDEN EAGLE	1906	275.7 ×32.1×10.1	793	30.5, 46.5, 71×60	N. 455
CRESTED EAGLE	1925	299.7 ×34.6×11.1	1,110	30, 46, 69.5×66	N. 538
ROYAL EAGLE	1932	289.75×36.5×11.5	1,538	28.5, 44, 68×60	I. 1300

ISLE OF ARRAN. 1892/1933.

In 1933, the Clyde Steamer ISLE OF ARRAN was bought from the Williamson-Buchanan Steamers, and made her inaugural trip on 5th June, 1933, from the Tower Pier to round the Nore Lightship.

The ISLE OF ARRAN ran trips to Herne Bay, on Sundays to Margate, and twice weekly a tour of the London Docks under the ægis of the Port of London Authority.

For some time her appearance was quite different from that of the EAGLES, the funnel being painted red with a black top, the lifeboats white, the hull grey—brown below the water line. With the commencement of the 1934 season, the funnel was cream colour with a black top, and the hull was painted black.

The ISLE OF ARRAN was built in 1892 by T. B. Seath & Co., Rutherglen, and formerly ran on the Glasgow-Arran service. She had a single cylinder—52 inches by 60 inches—engine of 161 n.h.p., steam being supplied by one " haystack " boiler. Dimensions, 210 feet long by 24.1 feet beam by 7.4 feet. Gross tonnage 313. At the close of the 1936 season she was sold to T. W. Ward & Co. for scrapping.

LAGUNA BELLE. 1896/1935.

In September, 1935, the LAGUNA BELLE was purchased, and in the following year was put on the Clacton and Walton service, on which she continued until the outbreak of War in September, 1939, when she was acquired by the Admiralty. (See The Belle Steamers, Chapter VII.)

ROYAL DAFFODIL. 1939.

This screw-driven vessel was built for the joint services of the G.S.N. Company and the New Medway Steam Packet Company, to operate between London and the Continent during the summer season, in conjunction with the QUEEN OF THE CHANNEL (see N.M.S.P. Co., Chapter VI) and the ROYAL SOVEREIGN.

The ROYAL DAFFODIL was launched in February, 1939,

P.S. " Isle of Arran " (1892)

Face page 100

T.S.M.V. " Royal Daffodil " (1939) *Photo: W. G. Tilling*

at Messrs. W. Denny Brothers' Leven Shipyard, Dumbarton. The motive power is provided by two sets of 12-cylinder Sulzer two-stroke marine type diesel engines developing 4,500 h.p., and giving a speed of 21 knots. Her dimensions are 299.7 feet in length, by 50.1 feet beam, and her gross tonnage 2,060.

The ROYAL DAFFODIL made her maiden trip to Calais on 28th April, 1939. She has a cruiser stern, two masts and two streamline funnels painted cream colour without the usual black tops; the hull was painted white, and to make her look more bright and cheerful for the passengers she had two green ribbon bands above and a green boot topping at the water line. On her spoon-shaped bow are the shield crests of the two Companies, which was also put on her funnels in 1947. A special feature of the ROYAL DAFFODIL, similar to the ROYAL SOVEREIGN though less exaggerated, is her flaring out amidships to form a species of dummy paddle sponson, which arrangement considerably increases her deck room.

No less than 2,000 passengers can be carried—all of one class—and no less than 286 can be accommodated in the two dining rooms at one sitting.

After the disappointing results of the KINGFISHER, the question may be asked why the introduction of this screw motor vessel, but it must be remembered that the vessel was built primarily for cross-Channel work.

This steamer perpetuated the name of the "Royal Daffodil" mentioned in the chapter dealing with the New Medway Company's fleet. The outbreak of war interrupted her first season. She was at once taken over by Sea Transport Service, and her war record, not less memorable than that of her predecessor, is worthy of note.

1939—Evacuated 4,000 women and children from London to Lowestoft. Conveyed part of the B.E.F. to France, via Southampton and Cherbourg.

1940—Transferred to the Fleet Air Arm. During the evacuations she made a voyage to Calais, the Channel Islands and six voyages to Dunkirk, during which time she brought home 9,500 troops. On her seventh trip she was bombed and badly damaged by enemy air attack, but, thanks to what has been described as outstanding seamanship and courage, was brought back to port.

1941-1945—Employed on the Larne-Stranraer military service.

1945-1947—Employed on the Dover-Calais, Newhaven-Dieppe and B.A.O.R. leave service until January, 1947, when she was de-requisitioned. During her war service this vessel carried a total of 2,443,979 service personnel, and steamed a distance of 170,000 miles. The services of her Master, Officers and crew were recognised by the awards of three D.S.C.s, one D.S.M. and three Mentions in Despatches.

After de-requisition she went back to her builders for re-conversion to her peace-time condition.

ROYAL SOVEREIGN. 1948.

On 17th July, 1948, there arrived in the Thames from the Clyde a new cross-Channel m.v. ROYAL SOVEREIGN, the third of that famous name on the London river. The new vessel, built by Messrs. W. Denny Bros., is intended for the cross-Channel service to Ostend, Calais and Boulogne when circumstances permit. This ROYAL SOVEREIGN has been built to replace the former vessel of the same name that was lost in the second world war, and is the last word in cross-Channel excursion steamers. Through the kindness of the owners, the author is enabled to illustrate the ROYAL SOVEREIGN. As will be seen she carries two masts and has a streamline funnel and is painted all white above the water line. The following are the principal particulars:—Length overall 285 ft., breadth moulded 48 ft., mean draft 8 ft. 9 in., trial speed 20.5 knots, gross tons 1,850. Passenger accommodation 1,783. The propelling machinery consists of two sets of 12-cylinder Sulzer engines each developing 2,250 b.h.p. The main deck has two dining saloons seating 96 and 140 diners respectively. The promenade has a smoke room, there is also a sun lounge 145 ft. by 33 ft. and a covered observation lounge 40 ft. in length. With the exception of the main engines everything is electrically operated including windlass, capstans and the steering gear. The ROYAL SOVEREIGN took up her duties on 24th July making her first trip to Margate and Ramsgate from the Tower Pier. Later in the season she ran on to Dover prior to making a cruise along the French Coast.

CRESTED EAGLE. 1938/1948.

A second CRESTED EAGLE now flies the G.S.N. Co. flag, this vessel being the former NEW ROYAL LADY. She was renamed primarily to retain possession of the name CRESTED EAGLE for a possible large Eagle steamer to be built in the near future. The present vessel was built in 1938 by J. Crown & Sons to the order of Mr. J. C. Round for making excursion trips from Scarborough. In 1947 she was running on the Forth. In 1948 she was purchased by the G.S.N. Co. for making the afternoon cruises from the Tower Pier to off Gravesend and alternatively the trips round the London Docks. The CRESTED EAGLE has a length of 144 ft. overall by 25.2 ft. by 9 ft. Mean loaded draft 5 ft. 4 in. Gross tonnage 248. Speed 13 knots. Motive power: Twin screw Diesel engines of 330 b.h.p. Passenger accommodation: 672.

Face page 104

T.S.M.V. " Royal Sovereign " (1937)

Photo: F. A. Plant

P.S. " Koh-i-noor " (1892)

CHAPTER V.

THE PALACE STEAMERS.

THE Palace Steamers comprised three of the most famous excursion steamers ever put on the Thames: the KOH-I-NOOR (1892), the ROYAL SOVEREIGN (1893) and the LA MARGUERITE (1894).

The KOH-I-NOOR was built to the order of the Victoria Steamboat Association, the ROYAL SOVEREIGN to the order of the London & East Coast Express Steamship Service, and LA MARGUERITE to the order of the Palace Steamers Ltd. All three were in the first instance run under the management of the Victoria Steamboat Association, though owned by separate Companies.

In April, 1895, the New Palace Steamers Ltd. was incorporated and under the management of this concern the three steamers continued to run until disposed of. Unfortunately the Palace Steamers Ltd. and the New Palace Steamers Ltd. were never able to pay the instalments due to the builders who were in effect the owners.

KOH-I-NOOR, 1892.

Built by the Fairfield Shipbuilding & Engineering Co., Govan. At the time, this vessel was by far the largest paddle steamer on the Thames, and also the fastest. Her cost was £50,000. On 19th May, 1892, she was officially tried on the measured mile at Skelmorlie, on the Clyde, covering the course four times. The mean speed developed was 19.499 knots.

The KOH-I-NOOR was built entirely of steel, and was 300.4 ft. long (310 o.a.) by 32.1 ft. beam (58 ft. over paddle-boxes) by 10.6 ft. from main deck. The engines were two-cylinder diagonal compound and placed forward of the paddle wheels. The cylinders, placed side by side, had flat or D-shaped valves, actuated by link motion, and were designed to indicate 3,500 H.P. The cranks were set at an angle of 90 degrees, the H.P. cylinder being 45 inches diameter and the L.P. cylinder 80 inches, with a 66-inch stroke.

Steam was supplied by four navy type boilers working at 120 lb. per sq. in. The paddle wheels were fitted with curved floats. The two funnels one forward and the other aft of the engines were made telescopic, so that they could be lowered to pass under London Bridge, as the passengers embarked at the Old Swan Pier, above the Bridge. The one mast was hinged for the similar purpose of clearing the Bridge.

The funnels were painted the standard V.S.A. colour: red, with black top and two white bands.

The KOH-I-NOOR was divided into water-tight compartments, which made her practically unsinkable. This division necessitated the first-class dining saloon being sub-divided into two. Two deck saloons, one at each end of the machinery space, supported an unbroken promenade nearly 300 feet in length and extending the full width of the vessel. She was fitted with a complete electric light installation, a post office, a hair-dressing establishment, two bathrooms, book and fruit stalls. The second-class saloon was of generous proportions.

Two rudders were fitted to the KOH-I-NOOR, one at the bow—in addition to the ordinary one aft—to facilitate the manœuvring of the vessel. The aft rudder, which was of large area, was nine feet long. The bow rudder, which took the form of the run of the ship, was completely within the stem.

When the KOH-I-NOOR was built, she left the Clyde for the Thames on 23rd May, 1892, but owing to the stress of weather put into Belfast Lough for two days. Proceeding thence on her voyage, she stranded on Porthgain Rock—off St. David's Head, on the coast of Pembroke—in a dense fog on the night of the 26th, causing considerable injury to the bow and a general straining throughout. After being backed off the rocks, she proceeded under her own steam with the tug ERNEST in attendance to Milford Haven where she was beached, later being put on the G.W.R. gridiron at Neyland. When the tide receded, the full extent of the damage was evident; and although the most serious damage was done to the bow, the injury was by no means limited thereto, as for a distance of about twenty feet from the bow the sides had collapsed like a concertina.

The KOH-I-NOOR was temporarily repaired and on leaving the gridiron was found to be perfectly tight. She then returned to the builders under her own steam. An entirely new bow was built and fitted to the vessel, which was finished and floated out of the dock in six working days an exceptionally smart piece of work.

The KOH-I-NOOR took up her running from the Old Swan pier to Southend and Clacton on 3rd July, 1892, and such a luxurious and handsome boat naturally met with instantaneous success. She was, however, soon put on the Thanet service, and for some years was the Saturday " Husbands' Boat " making two trips daily from Tilbury to Margate. After 1903 she ran to Dover.

When the 1914-18 War broke out, the Thames services naturally came to an end. The KOH-I-NOOR had gone to the Clyde just before to have new boilers installed, but subsequent events prevented the work being done. She was transferred to the Gareloch—an inland arm of the Firth of Clyde and a haven for vessels out of commission—and remained there until 1918 when she was sold to Messrs. Ward for £6,200. She was cut up at Morecambe in 1919.

ROYAL SOVEREIGN, 1893.

The ROYAL SOVEREIGN was the second of the Thames large excursion steamers. This popular vessel, also constructed and engined by the Fairfield Company, was launched on 17th April, 1893. She was built to the order of the London & East Coast Express Steamship Service, but as previously stated worked under the management of the Victoria Steamboat Association, who ran her to Margate, etc., from the Old Swan Pier.

In appearance, the ROYAL SOVEREIGN was almost identical with the KOH-I-NOOR, excepting the position of

P.S. " Royal Sovereign " (1893)

P.S. " La Marguerite " (1894)

Photo: Adamson, Rothesay

the funnels, the forward funnel of the new boat being further forward of the paddles than that of the earlier vessel. The two boats were almost identical in their fittings, accommodation, appointments and dimensions—300.2 by 33.1 by 10.6 ft., except that the KOH-I-NOOR had one foot more beam. The engines were similar in design and size, excepting an increase of three inches in the diameter of the L.P. cylinder, which was 83 inches, giving a slightly increased horsepower. The ROYAL SOVEREIGN had a deck bar forward of the funnels as well as aft; the KOH-I-NOOR had not.

In 1909, new boilers were fitted. During the war period she was laid up at Tilbury and in 1918 was sold to the Royal Sovereign Steamship Co. for the sum of £8,300. This Company intended to run her again as soon as conditions permitted. Difficulties arose, and another Company bearing the same name purchased her. Eventually, in March, 1929, she was sold to the G.S.N. Co. for £5,540, who ran her to Thanet during the 1929 season only, when she made 56 trips between London and Margate. During this season a black top was added to the cream-coloured funnels.

In February, 1930, she was sold for £3,500 to the Dutch shipbreaker T. G. Pas, Nieuw Lekkerland.

LA MARGUERITE, 1894.

The Palace Steamers in 1894 introduced a much larger and more luxurious paddle steamer named LA MARGUERITE after a daughter of Mr. Arnold Williams, a pioneer of improved steam service on the Thames; it was destined to bring London and Boulogne into closer communication.

The running of this vessel provided the first occasion on which the public could travel from London (Tilbury Pier) to Boulogne and back on the same day as a regular service, except for the South Eastern Railway service from Folkestone. LA MARGUERITE was the largest, fastest and most luxuriously fitted vessel to be employed on excursion traffic up to that time.

For ten seasons LA MARGUERITE was easily the most popular excursion steamer afloat, but unfortunately her popularity was not sufficient to make her running a profitable undertaking. It seems strange that the owners of such a popular boat should have decided to sell her after she had been running for ten seasons. If running costs were too high, one would have thought that she would have been sold sooner while she was worth more in the market. Eventually going to Liverpool, she was a commercial success there owing to the lower wages prevailing in Lancashire and the much lower price of coal.

The LA MARGUERITE was a 20-knot steamer built by the Fairfield Company, and like the KOH-I-NOOR and the ROYAL SOVEREIGN replete with every convenience and facility that the public could wish for. She was built expressly for the Boulogne service, calling at Margate. As she was too large and fast for the Thames Conservancy regulations, the point of departure was Tilbury. On one or two special occasions, however, she came up to Greenwich to make a trip to the Nore.

She was 350 feet b.p. by 40 by 13.6 from the main deck. The engines were two-cylinder diagonal compound, with cylinders 56 in., 110 in. by 77 in. the N.H.P. being 857 and the I.H.P. 7,500. As in the other two Palace steamers the engines were placed forward of the paddles, and at a casual glance gave the impression when in motion of going astern when actually going ahead.

The gross tonnage of these three steamers gives a good idea of their comparative size: KOH-I-NOOR 884 tons, ROYAL SOVEREIGN 891 tons, and LA MARGUERITE 1,554 tons. All three had their funnels painted red with a black top and two white bands.

In 1904 LA MARGUERITE was sold to the Liverpool & North Wales Steamship Co. who ran her on the Liverpool, Llandudno and Menai Bridge service, and she soon became as popular on the North Wales coast as she had been on the Thames. The funnels were then painted cream colour in accordance with the practice of the L. & N.W.S. Co.

In March, 1915, she was taken over by the Admiralty for use as a transport and remained as such until April, 1919, thus completing over four years on such work, chiefly between Southampton and various French ports. Her first trip in this capacity was on 17th March, 1915, when she embarked the 1st Battalion of the City of London Rifles. During her career as a transport she steamed over 52,000 miles—mostly during the night and without lights—and carried over 360,000 troops, the only mishap being a boiler explosion when four lives were lost.

For the summer service of 1919, LA MARGUERITE was chartered by the Isle of Man Steam Packet Co. for service between Liverpool and Douglas. In accordance with their practice the funnels were painted red with black tops. In the following year, she was again running on her old route from Liverpool to Llandudno and the Menai Straits. In the winter of 1925, this famous steamer was broken up at Briton Ferry by Messrs. T. W. Ward & Co.

LA BELGIQUE, 1875/1897, *ex* FLAMINGO, *ex* PARIS.

For one season only—1897—the New Palace Steamers Ltd. chartered the one-time cross-channel steamer PARIS and ran her between Tilbury and Ostend. This boat had an interesting career. Built in 1875 by J. Elder & Co., Glasgow (this firm later becoming the Fairfield Shipbuilding & Engineering Co.) for the L.B. & S.C. Ry. Newhaven-Dieppe service, she continued in that service until 1888. In that year she was repurchased by the builders who reconditioned her and installed new boilers. Two years later she was purchased by a Mr. Barnwell who ran her on the North Wales excursion traffic from Liverpool, under the name of FLAMINGO. In 1892, she was sold to Herr Ballin, the famous Hamburg shipowner, who traded her between Hamburg and Heligoland.

In 1895 she was bought back by Mr. Barnwell, and in 1897 was sold to a Mr. W. M. Rhodes, the latter chartering her to the New Palace Steamers who ran her under the name of LA BELGIQUE. After a long period of idleness she went north into the ownership of D. MacBrayne & Co., Glasgow, who ran her on the West Coast of Scotland as the GLENDALE. In 1905, this veteran was wrecked.

The LA BELGIQUE had a gross registered tonnage of 484, and was 220 ft. in length by 25.2 by 11; she was driven by compound oscillating engines, the H.P. cylinder being 41 inches in diameter and the L.P. cylinder 72 inches, with 60-inch stroke.

CHAPTER VI.

NEW MEDWAY STEAM PACKET COMPANY.

THE New Medway Steam Packet Company was incorporated in December, 1919, to acquire the assets of the Medway Steam Packet Company (incorporated in 1881) which was the successor to the Medway Steam Packet Company of 1837. The original Company was formed with the primary object of linking Chatham with Sheerness by water, prior to the advent of the railway.

It has not been possible to acquire any reliable informaton concerning the Company's boats prior to 1871, but it is of interest to record that the earliest of these boats were built of mahogany.

The New Medway Steam Packet Company are proprietors of their own shipyard, engineering shops, foundry, etc., and carry out all repairs and renovations with their own staff.

City of Rochester, 1847/1871.

The City of Rochester, like many of the Thames steamers, began its career in private ownership, being built at Chatham in 1849. In 1866, it is mentioned as being owned by a Mr. W. Giles of Rochester. About the year 1871, the vessel was acquired by the Medway Steam Packet Company. She was a small steamer, with a clipper bow, of only 73 tons gross register, dimensions being

P.S. " City of Rochester " (1847)

P.S. " Lady of Lorne " (1871)

123 feet by 14.5 by 7.7, with N.H.P. 40. She plied between Rochester, Chatham, Sheerness and Southend, and was broken up in 1897.

ALMA, 1855/ ?

The ALMA was built at Chatham in 1855 and was cut up in 1899. She was built of wood, and had the following dimensions: 118.4 ft. long by 12.7 by 6.6. Gross tonnage 69: N.H.P. 32. Was at one time owned by Mr. W. Giles, Chatham.

LADY OF LORNE, 1871.

The LADY OF LORNE was built in 1871 at Rochester, and survived until 1899. Dimensions: 116.7 ft. by 14.1 by 6.5. Registered gross tons 59; N.H.P. 36.

LADY MARGARET, 1883/1888.

The LADY MARGARET was acquired by the Company in 1888, having been purchased from the Bristol Channel Express Company. She had been built in 1883 by Russell's at Greenock, and was a big vessel in comparison with the previous boats. She was built of iron, and was 139.9 ft. long by 18.1 by 7.4. Gross tons 144; N.H.P.50. The engines were compound diagonal, by Allen & MacLellan of Glasgow, the cylinders being 22 in. and 37.5 in. by 30 in. She came to an untimely end in 1903, being destroyed by fire.

PRINCESS OF WALES, 1896.

The PRINCESS OF WALES was built in 1896 by R. Craggs & Sons, Middlesbrough. The engines were made by

Jameson & MacColl of Sunderland, and were of compound diagonal type with cylinders 21 in. and 39 in. by 30 in. She was first used to operate the Rochester-Southend service. In June, 1917, she was taken over by the Admiralty, renamed PADUA, and engaged on the Naval Medway Ferry service until released in April, 1920. In 1925, she was chartered by the Stanley-Butler S.S. Co., Kirkcaldy, and ran on the Firth of Forth. In March, 1927, she sank at her moorings but was raised and in due course returned to the Medway. In 1928, she was sold to Messrs. Ward for breaking-up. The PRINCESS OF WALES was built of steel; dimensions: 139.6 long by 18.1 by 7.5, with a gross tonnage of 163. N.H.P. of engines 65.

AUDREY, 1897/1922.

The AUDREY was built by Messrs. Armstrong, Whitworth & Co., Newcastle, in 1897, for the Tyne General Ferry Company. In 1910 she went to Ireland having been purchased by the City of Cork Steam Packet Company. In 1913 she was acquired by the Cork, Blackrock and Passage Railway to carry passengers and cargo between Monkstown, Queenstown and Aghada on weekdays and to run excursions on Sundays. In 1914 Mr. S. J. Shippick bought her, and she returned to England and ran from Easter, 1914, until the outbreak of war on excursion trips between Bournemouth, Swanage, Studland and Poole.

In common with many other paddle steamers, the AUDREY was taken over by the Government in 1915, and retained by the Admiralty until 1922, during which time she was engaged at Chatham on the Medway Ferry service. She was 126 ft. long by 21.1 by 8.0; Gross tons 203; N.H.P. 60. She had diagonal compound engines made by the Wallsend Slipway Co., with cylinders 17 in. and 34 in. by 42 in. Speed 12 knots. Passenger accommodation, 600.

After the cessation of hostilities, the AUDREY was acquired in July, 1922, by the N.M.S.P. Co., being principally employed on the ferry service between Southend and Sheerness. During her last season she was stationed at Ramsgate, making local coastal cruises from that port, after which she was sold, in September, 1929, to Messrs. T. W. Ward & Co. for about £500.

CITY OF ROCHESTER (II), 1904.

The CITY OF ROCHESTER (II) was by far the Company's largest steamer. Built in 1904 by J. Scott & Co., Kinghorn, Fife, she was of the following dimensions: 160.2 ft. by 22.2 by 7.3. Gross tons 235; N.H.P. 72. She had diagonal compound engines with cylinders 20 in. and 42 in. by 45 in. Her normal run was between Harwich, Felixstowe and Ipswich, and for a time to Herne Bay.

During the first World War, the CITY OF ROCHESTER (II) served as a Government tender at Chatham, from December, 1914, until November, 1919. She was one of the vessels that conveyed the men down the Medway to board the block ships waiting below the Nore for the memorable raid on Zeebrugge. On one occasion she had a remarkable escape from destruction. She had been lying alongside the PRINCESS IRENE in Sheerness harbour, and had only just left when the latter vessel was mysteriously blown up whilst lying at anchor. For a period she did a turn at mine-sweeping.

The foregoing boats were run by the old Medway Company, but subsequent boats were acquired by the New Company incorporated in 1919.

MEDWAY QUEEN, 1924.

In 1924, the Company commenced to advertise its vessels as "The Queen Line of Steamers," and in that year the Ailsa Shipbuilding Co., Troon, built the first "Queen," the MEDWAY QUEEN. Her tonnage is 318 gross, and she has a speed of 15 knots. She was the first of the Medway steamers to have her top deck carried forward to the bow of the vessel. At first the MEDWAY QUEEN was run between Rochester and Southend, but later was transferred to the Herne Bay service. Her dimensions are 179.9 feet by 24.2 by 7.8; N.H.P. 76; compound diagonal engines with cylinders 21 in. and 43 in. by 48 in. In 1938, she was reboilered and fitted for oil fuel burning.

QUEEN OF THE SOUTH, 1891/1924 (*ex* WOOLWICH BELLE).

Between 1924 and 1928 the Company added to its fleet five large paddle steamers, having purchased three ex-Belle steamers and two ex-Admiralty minesweepers. The first to be taken over was the WOOLWICH BELLE, which was purchased in 1924 from the Channel Excursion Steamers, Ltd., and then re-named QUEEN OF THE SOUTH. In 1931, she was laid by, and in July, 1932, was sold to London buyers for £300, and finally cut up at Grays.

ESSEX QUEEN, 1897/1925 (*ex* WALTON BELLE).

In 1925, the WALTON BELLE was purchased from the P.S.M. Syndicate, Ltd., and was re-named ESSEX QUEEN. In 1931, she was thoroughly re-conditioned, including new boilers, and equipped for burning oil fuel. Her principal services were to Clacton or Margate and Ramsgate, and frequently to London and the dock cruises. She was withdrawn from service in September, 1938, and sold in 1946 to the South Western Steam Navigation Co., Totnes, and re-named PRIDE OF DEVON.

QUEEN OF THANET, 1916/1927 (*ex* H.M.S. MELTON) and QUEEN OF KENT, 1916/1927 (*ex* H.M.S. ATHERSTONE).

In 1927, the Company purchased two very handsome ex-minesweepers of the " Ascot " or race-course class, the MELTON and the ATHERSTONE. Both had been built in 1916, the former by W. Hamilton & Co. of Port Glasgow, and the latter by the Ailsa Shipbuilding Company of Troon. The MELTON had originally been sold to Messrs. Hughes and Bolckow, shipbreakers, when the great naval scrapping programme was in force. The MELTON was re-named QUEEN OF THANET and the ATHERSTONE became QUEEN OF KENT. Their normal services were from Chatham, Southend, Margate to Calais, Boulogne or Dunkirk. In 1930, a regular service was run to Dover, but that does not appear to have come up to expectations, for it was not repeated in the following year. Commencing in the 1933 season, on every Wednesday a day trip was made from Clacton to Calais.

These two vessels with their cruiser sterns—which were the first and only paddle boats to have cruiser sterns (excepting of course the other paddle minesweepers of the "Ascot" class)—and the distance apart of their two funnels gave them a conspicuous appearance. Their dimensions were as follows:

QUEEN OF THANET

234.9 ft. by 29.1 by 9.2; Gross tons 792; I.H.P. 1,600.

QUEEN OF KENT

235.2 ft. by 29.1 by 9.2; Gross tons 798; I.H.P. 1,600

Both had compound diagonal engines: Cylinders 26.5 in. and 52 in. by 54 in. Prior to the 1931 season, both were re-conditioned, fitted with new paddle wheels, wireless installation, and equipped for burning oil fuel giving them an increased speed. They were withdrawn from service in September, 1938, but after war duties resumed their former work in 1946 with the main masts removed. In the summer of 1948 the QUEEN OF THANET was running between Ramsgate and Dover. Both vessels were sold in December, 1948, to the Southampton Steam Packet Co., and renamed SOLENT QUEEN and LORNA DOONE respectively.

QUEEN OF SOUTHEND, 1889/1928 (*ex* YARMOUTH BELLE).

The QUEEN OF SOUTHEND was purchased in 1928 from the East Anglia Steamship Co., who ceased to operate on the Thames excursions. This addition to the Medway fleet was the ex-Belle steamer YARMOUTH BELLE. Her service was generally Chatham-Felixstowe. In 1931, she was reconditioned similarly to the ESSEX QUEEN and fitted for burning oil. In 1936, the Company at their Rochester yard extended the promenade deck up to the bow:

P.S. " Lady Marguerite " (1883)

Face page 120

P.S. " Medway Queen " (1924)

P.S. " Essex Queen " (1897)

hitherto it had finished at the mast. In 1938, the name was changed to THAMES QUEEN, and in the spring of 1947 she was sold to Metal Industries Ltd. for scrap.

ROCHESTER QUEEN, 1906/1932 (*ex* GERTRUDE).

In 1932, the N.M.S.P. Co. purchased from the L.M. & S. Ry. their Tilbury-Gravesend ferry steamer GERTRUDE. This vessel, which was then named ROCHESTER QUEEN, unlike all previous Medway steamers, was screw-driven. She was built in 1906 by A. W. Robertson & Co., London, of steel and of the following dimensions: 125 feet by 26.7 by 8.5; Gross tons 255; N.H.P. 48. Being twin screw, she had two sets of compound engines with cylinders 12-in. and 24-in. by 18 in. She performed a double trip daily between Strood, Chatham, Sheerness and Southend. In 1931 she was withdrawn, and in 1933 was purchased by M. H. Bland & Co., Gibraltar, and re-named CAID.

CLACTON QUEEN, 1897/1933 (*ex* DUCHESS OF KENT).

The CLACTON QUEEN was purchased in November, 1933, from the Southern Railway Co. She was built in 1897, by Day, Summers & Co., Southampton, to the joint order of the L.B. & S.C. and the L. & S.W. Rys., for the Portsmouth-Ryde service. Her chief dimensions were: 195.4 feet by 26.1 by 9.0; Gross tonnage 399; compound diagonal engines with cylinders 32-in. and 59-in. by 60-in.; N.H.P. 225. In 1934, her original name DUCHESS OF KENT was changed to CLACTON QUEEN.

This steamer, although built for smooth water between the mainland and the Isle of Wight, was quite capable of holding her own in a bad sea when occasion arose. To give an instance, on 10th August, 1934, she set out from Clacton for a trip to Chatham, having on board 800 passengers. On the return journey she had to face a rough sea for ten hours. When she arrived off Clacton at 10.0 p.m., the sea was so rough that the captain thought it prudent to stand out to sea rather than try to disembark his passengers at the pier head. At 1.30 a.m. the next morning the CLACTON QUEEN had not arrived, although her lights could be seen from the shore, so the lifeboat put out and stood by for some time. Later as the gale subsided, an attempt to land the passengers proved successful.

In November, 1935, the CLACTON QUEEN was sold to a new Company designated the Mersey & Blackpool Shipping Company, Liverpool. It was the intention of the new owners to run a daily service between Liverpool and Blackpool, calling *en route* at New Brighton. The Steamer's name was changed to JUBILEE QUEEN, and the funnel painted buff with a black top and red band.

The Company was soon in difficulties, and the steamer again changed hands, the new owners being The Jubilee Queen Shipping Company. Commencing in August, 1936, the JUBILEE QUEEN inaugurated a new service between Barrow, Fleetwood and Blackpool. The red band round the funnel was removed. In May, 1937, she was sold to Messrs. Ward for £1,625, and cut up at Preston.

ROYAL DAFFODIL (I) (*ex* DAFFODIL) 1906/1934.

The ROYAL DAFFODIL was purchased by the N.M.S.P.

Co. from the Borough of Wallasey in 1934. This little twin-screw vessel (originally named Daffodil) that plied between Liverpool and Birkenhead, played a very prominent part at Zeebrugge on 23rd April, 1918.

H.M.S. Vindictive had been selected as the spearhead of the attack on the Mole, having the Daffodil and her sister boat Iris in attendance, all three carrying landing parties. On arrival, it was found that the Vindictive, on account of the heavy swell, had the greatest difficulty in coming alongside and in grappling the Mole to allow of the disembarkation of the landing parties. The Daffodil, following close astern, came up and in the most gallant manner placed her bow against the Vindictive and pushed her sideways until she was in correct position alongside. If it had not been for the continued pressure by the Daffodil towards the wall of the Mole throughout the operation, the return of the landing parties would have been impossible. The signal for the re-embarkation of the landing parties was given by the Daffodil's syren, as the syren and searchlight of the Vindictive had been shot away. The signal was made at 12.50 a.m., and had to be repeated many times on account of the excessive noise of the guns.

The whole action was one of those epic achievements which will live throughout the ages, and will ever redound to the glory of the little Daffodil, that she—together with the Iris—shared with the Vindictive the honours and ardours of the fight. The King commanded that for the rest of her career she should be known as Royal Daffodil. A metal plaque was affixed on board to commemorate the occasion. On her funnel and upperworks were to be seen the marks of the shrapnel which struck her during that historic engagement.

In normal service the Royal Daffodil worked between Strood and Sheerness. She was built in 1906 by Robert Stephenson & Co., Newcastle-on-Tyne, and was of the following dimensions: 152.1 ft. by 40.6 by 11.2; Gross registered tonnage 465. She was propelled by two sets of triple-expansion engines by W. Rollo & Sons, Liverpool. Cylinders, 16-in., 24-in. and 41-in. by 41-in. In April, 1938, the Royal Daffodil was bought by Van Heyghen Freres of Ghent, for about £1,000 and was broken up in Belgium. She therefore ended her days in the country where she earned her fame. Shortly before her withdrawal, she was under charter to the Port of London Authority and was used for cruises round the London Docks.

Queen of the Channel, 1935/1940.

In 1935, a revolution in Thames excursion steamers took place by the introduction of the motor-driven vessel. After the disappointing results of the G.S.N. Co's vessel Kingfisher—1907/1911—the introduction of a screw vessel was all the more remarkable.

There were many people intimately connected with the Thames excursion trade who maintained that it would always demand a paddler, and that any form of screw vessel was unsuitable for the conditions of the day trade. Their argument received added force by the fact that many motor ships are even more difficult to handle than direct-driven turbine vessels. The factor against this was that there were undoubted advantages in the diesel, if it could be contrived. It demanded, of course, smaller bunkers than any other type, and these of a kind that could be stowed away in odd corners. The crew would be smaller, which is a great consideration; further, there would be a great saving in running costs because the diesel engine can be shut down entirely when the vessel is not under weigh, whereas a steamer—no matter how carefully the fires are banked—is always using fuel, and even a minute stay alongside a pier will cause her safety valves to lift, which is a sure sign of waste. Against these points is the disadvantage already mentioned, of manœuvring, and the reputation which the diesel has for excessive vibration when put into comparatively small hulls where great speed is demanded.

These factors had to be weighed against one another and carefully considered, but Messrs. Denny, of Dumbarton, decided that the risk was well worth taking.

This pioneer ship, the QUEEN OF THE CHANNEL, was launched in May, 1935, and was ready for service by the end of June. She was originally owned by the London & Southend Continental Shipping Co., a concern specially formed for the purpose by Messrs. Denny and the N.M.S.P. Co., which later took over the vessel entirely. Her dimensions were 255 ft. by 34 by 11 ft. 6 in.; Gross registered tonnage 1,030. The twin screws were driven by two Sulzer two-stroke trunk-piston diesels, each having eight cylinders $14\frac{3}{16}$ in. by $23\frac{3}{8}$ in. One engine was built by Sulzer's in their own works in Switzerland, the other by Denny: both are of identical design, and develop 1,500 B.H.P. at about 320 R.P.M. In addition to the main engines there is a very elaborate auxiliary engine installation, particular care being taken to give the ship sufficient compressed air for the manœuvring which may be necessary when coming alongside the various piers.

With her hull painted white and two well-proportioned funnels (the forward one being a dummy used as a deck-chair store) she presented a handsome appearance. The passenger accommodation was no less than 1,600.

Most of the N.M.S.P.C. vessels are kept closely connected with Rochester, but during the first season the QUEEN OF THE CHANNEL was based on Gravesend. From Gravesend she crossed to Tilbury by arrangement with the L.M. & S. Ry., proceeding thence to Southend and Margate, before striking across Channel to Ostend, Calais or Boulogne, according to the day of the week. Designed for a speed of 19 knots, she averaged 21.9 on trial. Later, when the ROYAL SOVEREIGN was brought out as an improvement on her design, she transferred to the East Coast trips.

The Cunarder, QUEEN MARY, sailed on her maiden trip from Southampton to New York on 27th May, 1936. To witness that important inaugural departure, the N.M.S.P. Co. ran the QUEEN OF THE CHANNEL on a two-day excursion from Southend and Margate, leaving on 26th, making a call at Brighton, and staying overnight.

The author, on this occasion, witnessed an incident which happily did not turn out to be serious. The vessel came along the coast and then turned in for the landing stage on the west side of the Brighton West Pier. The pier staff were ready to make fast when to everyone's surprise she shot past and went straight for the beach, stopping almost broadside on. However, after a short while, she went astern and got off, backing to the landing stage none the worse for her too close visit to the Queen of Watering Places. On the following day the trip was resumed to Spithead and back to Margate and Southend.

This handsome vessel was sunk during the Dunkirk evacuation, May, 1940, but a new Queen of the Channel is under construction for the 1949 season.

Royal Sovereign, 1937.

Owing to the success of the M.V. Queen of the Channel, the N.M.S.P. Co. decided to have another vessel of similar design but of more commodious dimensions. Messrs. Denny, as before, were the builders, and launched the vessel in April, 1937. The new ship was named Royal Sovereign, and was delivered in time for the summer traffic. Ostend was the objective, and the vessel was scheduled to leave the " Pool " at 8.40 a.m. calling at Southend *en route* at 11.30 a.m. On the return, passengers left the Belgian resort at 8.0 p.m. for Tilbury, and completed the last lap of the journey to London by special train.

The Royal Sovereign was a big improvement on the Queen of the Channel, being 269.6 feet in length by 47.0 by 9.1, and of 1,527 tons gross. She was propelled by twin screws driven by two sets of 12-cylinder Sulzer engines which gave a speed of over 21 knots. Running alongside the hull she had the form of a paddlebox sponson, which allowed greater space in the main lounge and dining saloon, and also added considerably to steadiness at sea. The sponsons (or " flaring-out ") were over 150 feet in length, and extended from a point well forward of the bridge to the quarter, the main and promenade decks, being extended over them. The ROYAL SOVEREIGN carried two masts but only one funnel.

In 1937, the General Steam Navigation Company acquired the share capital of the New Medway Steam Packet Company, which concern is now controlled by the former. The amalgamation has not affected the services, as the two concerns continue to operate separately: the advertised " Eagle " steamers of the G.S.N. Co. run between London, Margate, Ramsgate and up the Essex coast to Clacton, whilst the N.M. Co.—popularly known as the " Queen Line "—continue to serve the Medway piers, Gravesend, etc., to the Continent.

Similarly to the other motor vessels and ROYAL EAGLE, the ROYAL SOVEREIGN's funnel was coloured buff minus a black top, whereas all the other excursion steamers have a black top to the funnel.

When the ROYAL SOVEREIGN was ordered by the N.M.S.P. Co., the name chosen was CONTINENTAL QUEEN, but was altered to the present name when the G.S.N. Co. acquired control.

The War career of the ROYAL SOVEREIGN was somewhat brief. Immediately prior to the outbreak of war she took part in the evacuation of women and children from London to the East Coast, and later acted as a troopship

P.S. " Queen of Kent " (1916) leaving Gorleston

Photo: Martley Burtt

Face page 128

T.S.M.V. " Royal Sovereign " (1948)

between Southampton and Cherbourg for the B.E.F. She took part in the Dunkirk evacuation in 1940, but unfortunately was sunk by a mine in the Bristol Channel on the 9th December, 1940.

Rochester Queen, 1947.

In 1947, the New Medway Steam Packet Co. purchased a twin-screw former military landing craft for the purpose of converting it into a small excursion vessel to run between Sheerness and Herne Bay. She went into service in 1948 carrying the name Rochester Queen: the second vessel to bear that name. She has a smart appearance with white painted hull and cream funnel cruiser stern and a raking bow which gives an impression of speed. The Rochester Queen was built as recently as 1944 at Stockton-on-Tees by the Stockton Construction Co., her registered dimensions are: Gross registered tons 345, length 147 ft., beam 22.5 ft., depth 10 ft. Motive power two 4-stroke single-acting geared Diesel engines by Davey, Paxman & Co., Colchester. There are 24 cylinders 7 in. diameter and 7¾ in. stroke. On trials the Rochester Queen's speed was 12 knots with a draught of 6 ft., and carrying 425 passengers. The passenger accommodation consists of a promenade deck, fore deck and after main deck. There is also a dining saloon, lounge and refreshment bar for passengers convenience.

CHAPTER VII.

THE BELLE STEAMERS.

As stated in Chapter III, the River Thames Steamboat Co. was financially in low water and in 1887, owing to the pier dues not being paid, the Clacton Pier Company refused to let that Company's passengers land there. The service was, therefore, stopped. It was stated that 25,000 passengers landed there each season and the stoppage of the boats seriously affected the popularity of the town as a holiday resort. Some of the more influential people who were developing the place—especially the shareholders of the Clacton Pier Company—therefore decided to run a steamship of their own and formed the London, Woolwich & Clacton-on-Sea Steamship Company and had built a paddle steamer named CLACTON by the firm of J. Scott & Co. of Kinghorn.

This vessel was built of steel, 189.5 ft. in length by 22.2 by 8.3, and of 241 tons gross and was fitted with compound engines having cylinders 28 and 57 in. diam. by 57 in. stroke of pistons. The N.H.P. was 204. She was a great improvement on the River Thames Company boats and had a large saloon 40 ft. long extending the whole breadth of the vessel. Beneath this was a fine dining saloon and the quarterdeck formed a shaded retreat under an awning. She developed a speed of 16 knots on the measured mile. To facilitate passing under London

Bridge, the CLACTON had a telescopic funnel and a hinged mast but the rake of the mast was different from that of the funnel, which somewhat spoiled the symmetry of the boat's appearance.

The CLACTON was put into service on May 17th, 1888, and was scheduled to leave the Old Swan Pier daily (except Fridays) at 9.30 a.m., arriving at Clacton at 2.30 p.m. and arriving back at London at 8.30 p.m.

As mentioned in Chapter IV, the General Steam increased their fleet by five new boats in 1887-1889 and had, therefore, boats available for the Clacton run, and the CLACTON, after running only one season (1888), was sold and in the following year was running out of Constantinople under the name AIDIN. The AIDIN was one of the first Turkish steamers to carry anything like 110 lbs. of steam and in consequence gave a good deal of trouble to the Turkish engineers. For a time an English engineer in the employ of the Turkish Government was placed on board. The Bosphorus boats were notorious for overcrowding and the AIDIN, when carrying a full load, generally had a bad list with her sponsons touching the water. However, she remained in service until 1913.

Presumably dissatisfied with the service given Clacton by the G.S.N. Co., the Company decided to run their own steamers again and during the next six years Denny Bros. of Dumbarton built four vessels; CLACTON BELLE (1890), WOOLWICH BELLE (1891), LONDON BELLE (1893) and SOUTHEND BELLE (1896).

In 1897, the Company was reconstructed as Belle Steamers Ltd., and at the same time it extended its services along the Kent coast to Herne Bay and Ramsgate, ordering from Denny's two more steamers; WALTON BELLE (1897) and YARMOUTH BELLE (1898).

In 1898, the Company was again reconstructed as the Coast Development Corporation, Ltd., and the seventh and last " Belle "—SOUTHWOLD BELLE—appeared in 1900. All seven vessels were the product of Denny's of Dumbarton.

The Belle steamers took a great part in attracting Londoners on holiday bent from the railways to the sea and at this period the Thames enjoyed almost as great facilities as the Clyde, which has always been favoured in this respect.

All services ceased immediately on the outbreak of the first World War.

All the Belle boats were of different dimensions, the largest being the LONDON BELLE, which was the first of the fleet to have triple expansion engines, a practice that was followed in the four subsequent boats. They differed in their saloon arrangements, for instance, the WOOLWICH BELLE had no deck saloon forward of the paddle-boxes, the SOUTHWOLD BELLE had a fore-deck saloon the full width of the vessel. The CLACTON BELLE originally had no fore-deck saloon but later one was added. She had a telescopic funnel when built which was painted all black. The aft saloon was longer and had more windows which distinguished her from the other Belles.

All the vessels were modelled on similar lines, the distinguishing features were that they had only one mast and one buff or cream coloured funnel. The navigating bridge was placed immediately behind the funnel, and the saloon or upper promenade deck did not extend beyond the mast. By these features it was quite easy to identify a Belle steamer from that of any other company, but it required a much more intimate acquaintance with the boats to detect one Belle steamer from another. There were, however, certain details in which they differed, and by which they could be picked out from a distance, the principal of which were as follows:—The LONDON BELLE had a much taller funnel than on the other boats. The SOUTHEND BELLE had a shorter and broader funnel than the LONDON BELLE and carried a large circular ventilator just aft of the bridge. She also had a support for a light on the after rail of the bridge instead of on the funnel as on other boats.

The CLACTON BELLE had a tall funnel in two steps and her mast was much shorter than those of the rest of the fleet. The WALTON BELLE, YARMOUTH BELLE and SOUTHWOLD BELLE each had a short, broad funnel having a square base, and it was difficult to discern one from the other, but it could be noted that the WALTON BELLE carried fewer ventilators, on account of the fact that she was the smallest of the fleet, the WOOLWICH BELLE could be easily identified by size alone.

With the exception of the SOUTHWOLD BELLE, YARMOUTH BELLE and the WALTON BELLE the steam whistles were placed in front of the funnel and the exhaust pipes behind. All the boats were fitted with steel floats to the paddle wheels with the exception of the CLACTON BELLE which had wooden floats. For handiness in navigating the upper Thames a bow rudder and suitable steering gear was provided.

CLACTON BELLE.

Built 1890. Dimensions:—Length 246 ft. by 26.5 by 9.5. Gross tons 458. Compound diagonal engines, cylinders 22 in., 50 in. by 60 in. N.H.P. 161. Speed 17 knots. Two boilers. Sold 1912 to the P.S.M. Syndicate Ltd. In 1922 purchased by the Royal Sovereign Steamship Co., three years later sold to the East Anglia Steamship Co. In 1929 sold to Messrs. T. W. Ward and cut up at Grays.

WOOLWICH BELLE.

Built 1891. Dimensions:—Length 200 ft. by 24.1 by 7.6. Gross tons 298. Compound diagonal engines, cylinders 24 in., 42 in. by 54 in. N.H.P. 109. Speed 15 knots. One boiler. She was the smallest of the Belle fleet and was employed on a service between Harwich and Ipswich for a time. In 1922 purchased by the Channel Excursion Steamers Ltd. who renamed her QUEEN OF THE SOUTH and during the summer seasons of 1922 and 1923 she was engaged in running trips from Brighton. In 1924 she was purchased by the New Medway Steam Packet Co. In 1931 she was laid up and in July 1932 sold for £300, being finally cut up at Grays.

LONDON BELLE.

Built 1893. Dimensions:—Length 280 ft. by 30.0 by 10.9. Gross tons 738. Triple expansion engines, cylinders 29.5 in., 44 in., 64 in. by 72 in. N.H.P. 348. Speed 19

P.S. " Clacton " (1888) as built

P.S. " Clacton " as running in Turkish waters named " Aidin "

Face page 134

P.S. " Laguna Belle " (1896)

knots. Two boilers. Sold in 1921 to the P.S.M. Syndicate. In 1922 leased to the Royal Sovereign Steamship Co., and in 1924 purchased by that concern. In 1929 cut up at Grays by Messrs. Ward.

SOUTHEND BELLE.

Built 1896. Dimensions:—Length 249 ft. by 30 by 10. Gross tons 570. Triple expansion engines, cylinders 28 in., 41.5 in., 60 in. by 60 in. N.H.P. 570. Speed 18 knots. Two boilers. Sold 1921 to the P.S.M. Syndicate. In 1923 disposed of to the R.S.S. Co. and in 1926 the East Anglia Steamship Co. was registered as the owners. In 1929 was transferred to the East Anglian Hotels Ltd. who renamed her LAGUNA BELLE. In June, 1931, she came back to the Thames and ran daily from the Tower of London pier to Clacton calling at Greenwich and North Woolwich. The LAGUNA BELLE was distinguished by a blue band round the funnel and a blue and white triangle shaped sign at the bows.

In September, 1935, she finished her London-Clacton season and was purchased by the G.S.N. Co., who ran her to Clacton and Walton until the outbreak of war, in September, 1939, when she was taken over by the Admiralty. The LAGUNA BELLE was the only steamboat to run to Clacton direct, all the other vessels calling at Southend. The mid-week return fare to Clacton was extremely cheap, at one time being as low as 4s. For the Clacton service the fore-saloon was re-constructed and plated in and a complete amusement saloon was provided. After purchase by the General Steam the LAGUNA BELLE was painted in their standard pleasure steamer colours, but the black band at the top of the funnel was much deeper than on their other vessels.

WALTON BELLE.

Built 1897. Dimensions:—Length 230 ft. by 26.2 by 9.2. Gross tons 465. Triple expansion engines, cylinders 20.5 in., 30 in., 43 in. by 60 in. N.H.P. 143. Speed 17 knots. One boiler. Sold to the P.S.M. Syndicate in 1921, in 1923 to the R.S.S. Co. and in 1925 purchased by the New Medway Steam Packet Co. and renamed ESSEX QUEEN.

YARMOUTH BELLE.

Built 1898. Dimensions:—Length 240 ft. by 28.1 by 9.3. Gross tons 522. Triple expansion engines, cylinders 20.5 in., 30 in., 43 in. by 60 in. N.H.P. 143. Speed 17 knots. One boiler. In 1921 sold to the P.S.M. Syndicate, but in 1923 the R.S.S. Co. were registered as the owners and in 1926 the East Anglia Steamship Co. In 1928 was acquired by the New Medway Steam Packet Co. and then was renamed QUEEN OF SOUTHEND and in 1938 renamed THAMES QUEEN. After the 1935 season considerable alterations were made, the original open fore-deck being built up, thus enabling the promenade deck to be extended to the bows.

SOUTHWOLD BELLE.

Built 1900. Dimensions:—Length 245.3 ft. by 28 by 9.3. Gross tons 535. Triple expansion engines, cylinders 20.5 in., 30 in. and 43 in. by 60 in. N.H.P. 143. She had two boilers. In 1912 purchased by the Hamburg-Amerika Line who renamed her WESTERLAND, her port of registry being Hamburg. In the year following she went to Cherbourg and was re-named BON VOYAGE. In 1922 she was registered at Genoa, her owners being the Societa Anonima Pragma. In 1924 she was transferred to the firm of Vasala & Narizzano who broke her up at Genoa in 1925.

* * * *

With the outbreak of the first World War the Germans started sowing mines in large numbers and the Admiralty requisitioned a large number of steam trawlers to sweep them up. Many of the vessels were lost and the Admiralty then decided to requisition all the suitable pleasure steamers that lay idle in various ports.

The YARMOUTH and CLACTON BELLES were taken over in August, 1915, the WALTON BELLE in the following December, the LONDON BELLE in March, 1916, and the SOUTHEND BELLE in April, 1916.

The conversion and fitting out of each steamer extended over a period of about six weeks. The first steamer ready for work was the YARMOUTH BELLE and with the others that followed all were put to work sweeping in the neighbourhood of the Thames; Sheerness being their base. Later, they were shifted to Harwich and remained in that vicinity until August, 1917. On one occasion the YARMOUTH BELLE was paired with the Caledonian Steam Packet Company's DUCHESS OF HAMILTON when the latter was blown up by a mine. The YARMOUTH

BELLE, though damaged by the explosion, succeeded in saving some of her partner's crew.

The LONDON BELLE was responsible for saving the passengers and crew of the G.S.N. Co's. PEREGRINE which foundered off the Sunk lightship in December, 1917. The CLACTON BELLE came into the news when she took on board the crew of the German airship L19 which came down in the Thames estuary when returning from a bombing raid on Felixstowe in 1917. The airship was taken in tow to Sheerness. One of the duties of the " Belles " was to sweep in front of warships and during these operations their trips extended to the French and Danish coasts. Towards the latter part of 1917, when Liverpool was used as a port for landing American troops, and enemy mines were then being laid in the vicinity, four of the " Belles " were sent round to the Irish Sea having Liverpool as a base and remained their until withdrawn in 1919.

The LONDON BELLE and WALTON BELLE on their return from mine sweeping early in 1919 were promptly dismantled of their mine-sweeping gear and fitted out for further urgent Government service as hospital tenders. There names were removed and they were known as H.C.2* and H.C.3 respectively. They were fitted up with a large number of cots in the two first-class saloons, an operating room abaft the main saloon, mess room and cabins on the saloon deck for surgeons and officers, extra lavatory, heating and ventilating apparatus, and an awning from the bridge to the sponson rails.

* H.C. = hospital carrier.

The fitting out was finished in May, 1919, when they at once proceeded to the White Sea, a distance of over 2,000 miles from the Thames, where they were intended to be used for the conveyance of wounded men from the Russian campaign down the Dvina to the sea for transfer to ships that were to convey them home. The trip took about 20 days each way, making stops at Lerwick, Bergen and Hammerfest. During the trip some very heavy seas were encountered in crossing the North Sea and rounding the North Cape but nevertheless the brave little paddlers came through without any mishap.

Very little hospital work was carried out, consequently the vessels were used for other purposes. H.C.2 and H.C.3 came under fire on several occasions and suffered a number of casualties amongst the landing parties and crews. Many impressions from machine guns and snipers were made on the funnels, ventilators and other parts of the ships. After working for about four months in the Arctic Seas around Archangel, H.C.2 and 3 returned to the Thames having completed over four years service unequalled by any other boats of this description. It was a great tribute to the "Belle" steamers that out of the large fleet of similar boats engaged, they alone as a fleet came through without loss, and it is doubtful if any line of steamboats could show a finer record of work. They were released by the Admiralty as follows:—CLACTON BELLE and SOUTHEND BELLE, November, 1919; YARMOUTH BELLE, January, 1920; WALTON BELLE, May, 1920 (H.C.3); and LONDON BELLE, July, 1920 (H.C.2).

CHAPTER VIII.

LONDON COUNTY COUNCIL AND LATER URBAN SERVICES.

THE Thames is a fine river and eminently suitable to sustain a commercial passenger service but it suffers from several great disadvantages. It is far from straight and does not provide the shortest route between two places at any distance apart; the main thoroughfares running east and west do not touch its course; there are heavy pier dues to be paid, and it would have to compete with excellent road and underground facilities. In the outcome, it is used only for pleasure trips, the enjoyment of which is dependent on the fickle English weather.

When early in 1902 the Thames Steamboat Company announced that their service would be discontinued, the London County Council immediately promoted a Bill in Parliament to empower them to run a service, but the Bill was thrown out.

Realising the relief it would give to road traffic congestion, the Council again applied for the necessary powers. These were obtained in 1904 and a service was started on 17th June, 1905, between Hammersmith and Greenwich with a fleet of thirty steamboats; H.R.H. The Prince of Wales inaugurating the service on the KING ALFRED. The boats were worked at such a heavy loss that they were taken off in October, 1907, and on December 15th, 1907, the Council decided to sell the thirty boats and invited tenders for their purchase. The total deficiency for the three years' working was £162,499.

The L.C.C. disposed of fourteen boats to a new concern called the City Steam-boat Company which worked a ten minute service from 1909 until August, 1914, when the service stopped with the outbreak of war. The other boats were distributed far and wide.

When these boats were built they were criticised by some people because the designers did not copy the screw principle of the Paris " fly boats," although it is obvious that the fact that there was only about three feet of water at low tide at many spots between Vauxhall and Putney made it necessary to keep their draught down to 2 ft. 10 in. which could only be done by paddles.

The boats cost rather more than £6,000 each, the original contracts awarded 10 boats to Thornycroft's, 10 to Napier & Miller and 10 to the Thames Iron Works. As they were wanted in a hurry, some of the boats were sub-contracted to other builders. The boats were practically all of the same dimensions: 130 ft. in length by 18 by 6.8, 116-125 gross tons. They all had diagonal compound engines having cylinders 16 in., 31 in. by 36 in., I.H.P. 350. Steam was supplied by one boiler. Diameter of paddle wheels 10 ft. 6 in. Speed on trials 12.5 knots. Carrying capacity 530 passengers.

They ran throughout the year to a time-table irrespective of the tides, starting from Greenwich and Hammersmith at 7 a.m. and then leaving every 15 minutes until 6.30 p.m. calling at 23 piers. There were later boats up to 7.15 p.m. from Hammersmith and Greenwich to Westminster. On Sundays they began two hours later. In addition there were workmen's boats at 6.30 a.m. and express boats in the mornings and evenings of weekdays which did the journey from Hammersmith to London Bridge in 70 minutes and between Greenwich and Westminster in 45 minutes. The fares were based on penny stages: a single ticket for the whole journey cost 5d. and a return 8d., workmen's fares being 1d. single, 2d. return.

They were built as follows:—

ALLEYN, BOYDELL, BRUNEL, CARLYLE, GIBBON, GRESHAM, KING ALFRED, MORRIS, PURCELL, SLOANE, and VANBRUGH by the Thames Iron Works.

BEN JONSON, FRANCIS DRAKE, RALEIGH, SHAKESPEARE by J. I. Thornycroft & Co., Woolston.

CAXTON, CHARLES LAMB (originally intended to be BAYNARD), CHAUCER, COLECHURCH, EARL GODWIN, EDMUND IRONSIDE, FITZAILWIN, OLAF, THOMAS MORE, TURNER, and WHITTINGTON by Napier & Miller & Co., Glasgow.

CHRISTOPHER WREN, MARLOWE, PEPYS, and RENNIE by G. Rennie & Co., Greenwich

The subsequent disposal of the boats and the amount their sale realised was as follows:—

ALLEYN	20/7/09	£705	To Dundee and subsequent owners. Finally to Mesopotamia.
BOYDELL	15/6/09	£990	To Royal Servian Privilege Steamship Co.
BRUNEL	20/7/09	£500	To J. Parson, Plymouth.

Ben Jonson	20/7/09	£500	To Dampfschiffgesellschaft der Vierwaldstättersee. When purchased by the Swiss the boat steamed across the Channel up the Rhine to Basle. Taken to pieces, put on rail and re-assembled at Lucerne. Renamed Rhin.
Carlyle	15/6/09	£990	To Tay Steamboat Co., Dundee.
Caxton	do.	£975	To C. A. Stewart & Co., London.
Charles Lamb	/09	£393	To City Steamboat Co. Resold to Genoa renamed Zulema.
Chaucer	do.	do	To City Steamboat Co. Resold to Nantes and renamed Ville D'Angers.
Christopher Wren	do.	do.	To City Steamboat Co. Finally to Mesopotamia.
Colechurch	20/7/09	£500	To Lake Lugano.
Earl Godwin	/09	£393	To City Steamboat Co. Resold to Dumbarton & Balloch Joint Line Committee. Renamed Queen Mary.
Edmund Ironside	/09	£393	To City Steamboat Co. Finally to Mesopotamia.
Francis Drake	6/4/09	£1,012	To Escaut & Rupel Steam Navigation Co., Tamise, Belgium. Renamed Wilford I.
Fitzailwin	/09	£393	To City Steamboat Co. Finally to Mesopotamia.
Gibbon	6/4/09	£1,000	To K. Bohre', Esschenweg, Rotterdam. Resold to Mafalda Soc. di Nav. a Vao. Puglia, Bari, Italy. Renamed Mafalda.
Gresham	/09	£393	To City Steamboat Co. Resold and went to the Rhine.
King Alfred	11/5/09	£1,050	To Glaser, Bewer & Co., London. Resold to O. May, Memel, for service on the Rhine.
Marlowe	/09	£393	To City Steamboat Co. Resold to Germany for service between Cologne and Mulheim.
Morris	do.	do.	do. do.
Olaf	20/7/09	£500	To Lake Lugano.
Pepys	/09	£393	To City Steamboat Co. Resold to Cie. Lyonnaise de Nav. et Remorquage, Lyons. Renamed Ville de Valence.
Purcell	15/6/09	£980	To Royal Servian Privilege Steamship Co.

RALEIGH	/09	£393	To City Steamboat Co. Resold to Soc. de Suc. et Raf. de Ponts, Belgium.
RENNIE	do.	do.	do. do.
SHAKESPEARE	do.	do.	To City Steamboat Co. Resold to Dumbarton & Balloch Joint Line Committee. Renamed PRINCESS PATRICIA. Sold at Balloch in 1938 and cut up.
SLOANE	do.	do.	To City Steamboat Co. Resold to Germany for service on the Rhine.
THOMAS MORE	6/4/09	£1,012	To Escaut & Rupel Steam Navigation Co., Tamise, Belgium. Renamed WILFORD II.
TURNER	20/7/09	£500	To Germany for service on the Rhine.
VANBRUGH	15/6/09	£990	To Boieldieu Cie Rouennaise de Nav., Rouen.
WHITTINGTON	20/7/09	£500	Went to North of Russia.

The engines of the L.C.C. steamboats were made by Scott's Engineering & Shipbuilding Co., Greenock, excepting those for the ALLEYN, BOYDELL, CARLYLE, GIBBON, KING ALFRED, PURCELL, SLOANE and VANBRUGH which were made by the Thames Iron Works Co.

Since the withdrawal of the London County Council steamers in 1907 there has not been an all-the-year round service in operation. It has always been recognised that such a service would help relieve the ever-growing congestion of London's road traffic, and from time to time schemes have been put forward.

In 1934, a committee designated the Thames Waterbus Committee, submitted two schemes to the Ministry of Transport. One proposed a fleet of fifty boats to work a four-minute service between Hammersmith and London Bridge and a six-minute service between London Bridge and Woolwich during the rush hours, with various inter-

P.S. " Earl Godwin " (1905) *Courtesy of L.C.C.*

Face page 144

M.V. " Oat " (1948)

Courtesy of "Motor Boat & Yachting"

mediate stops. A company entitled Thames Passenger Services Ltd. with a capital of £500,000 was proposed, the interest to be guaranteed by the Treasury or by the London Passenger Transport Board. It was estimated that it would be necessary to carry 21½ million passengers a year at an average fare of 2½d. to make the service remunerative. The second scheme, submitted by Thames Taxi's and Waterbuses Ltd., provided for a capital of £300,000, without Treasury guarantee; this company based its figures on carrying 3½ million passengers a year at an average fare of a shilling. The London Passenger Transport Board decided that their financial commitments did not allow them to support either scheme, and as financial success was considered doubtful nothing further was done.

During the early part of the second world war a service was started to help relieve the traffic congestion that arose at that time. Six up-river boats were obtained from the firm of J. Mears, Ltd. of Richmond, and from 13th September, 1940, they worked between North Woolwich and Westminster, calling at Brunswick, Greenwich, West India Dock, Tunnel, Cherry Garden and Tower piers. The first boat started at 6.30 a.m. in each direction to cater for the early morning workers. This service closed down on November 11th, 1940.

The latest venture came into operation on 27th July, 1948, when the towage firm of R. G. Odell Ltd. of Shepperton, started a half-hourly service between Putney and Greenwich. There were eight boats in commission named OAT, OBE, OCENIA, ODELIA, OENID, OFFICER,

OGANIC and OHMED. All were built by the Walton Yacht Works Ltd. during 1947/8 and are of the same design, round-bilged, with a length of 66 ft., breadth 15 ft., depth 5 ft. 6 in., and a draught of 3 ft. 6 in. but with alternative layouts.* The displacement with full load of 112 passengers is 24 tons. They are fitted with 6-cylinder, 60 h.p. Ruston-Diesel, Crossley and Perkins engines giving a service speed of 10 knots. The passenger accommodation comprises two large saloons which have an easily operated sunshine roof. When the service ceased on October 31st no less than 468,580 passengers had been carried. A new design is now under construction for 1949.

THAMES LAUNCHES LTD.

In May, 1945, the fleet of steam and motor up-river boats that had been operated for many years by J. Mears & Co., Richmond, were taken over by Thames Launches Ltd. The fleet consists of 17 boats, of which six are steam propelled, nine have Diesel motors and two are petrol driven. During the summer months the craft operate between Westminster and Hampton Court.

STEAM BOATS.

Name	Built	Where built	Length	Beam	Depth	G. Tons
THE KING	1902	E. Molesey	81.0ft.	14.6ft.	5.5ft.	50
VISCOUNT	1908	Oxford	100.8ft.	16.6ft.	4.6ft.	69
CONNAUGHT	1911	do	106.0ft.	16.6ft.	4.4ft.	78
KINGWOOD	1915	do.	101.2ft.	16.5ft.	5.0ft.	79
QUEEN ELIZABETH	1924	do.	110.0ft.	16.7ft.	5.0ft.	91
ABERCORN	1925	do.	110.0ft.	16.7ft.	5.0ft.	92

The last three boats have cruiser sterns.

* Presumably the river has been dredged since the L.C.C. venture to allow the increased draught.

MOTOR BOATS

Name	Built	Where built	Length	Beam	Depth	G. Tons
PRINCESS BEATRICE	1896	Hampton	74.9ft.	14.4ft.	5.9ft.	52
PRINCESS MARY	1904	E. Molesey	60.0ft.	11.1ft.	4.3ft.	25
HIS MAJESTY	1906	Maidenhead	84.1ft.	14.7ft.	5.9ft.	53
ROYALTY	1913	Oxford	101.1ft.	16.5ft.	5.1ft.	79
HURLINGHAM	1915	do.	101.2ft.	16.5ft.	5.0ft.	79
MARCHIONESS	1923	do.	85.5ft.	14.5ft.	4.3ft.	46
VISCOUNTESS	1926	do.	110.3ft.	16.6ft.	4.7ft.	83
KINGSTON	?	?	42.0ft.	7.0ft.	3.6ft.	?
TWICKENHAM	?	River Nene	36.0ft.	9.6ft.	3.6ft.	6

PRINCESS BEATRICE.	Converted	1946-7	from steam to	"Atlantic"	Diesel.
PRINCESS MARY.	do.	1946	do.	do.	do.
HIS MAJESTY	do.	1947	do.	"Crossley"	do.
ROYALTY	do.	1948	do.	do.	do.
HURLINGHAM	Converted	1947	from steam to	"Crossley"	Diesel.
MARCHIONESS	do.	1947-8	do.	do.	do.
VISCOUNTESS	do.	1945	do.	do.	do.

KINGSTON — Originally named HALCYON.

TWICKENHAM — Converted from steam to "Perkins" Diesel. Originally named QUEEN OF NENE.

PETROL-PARAFFIN BOATS.

Name	Built	Where built	Length	Beam	Depth	G. Tons
MUTT	1926	Eel Pie Island	40.0ft.	12.6ft.	3.6ft.	7
JEFF	1926	do.	40.0ft.	12.6ft.	3.6ft.	7

CHAPTER IX.

THE THAMES FERRY SERVICES.

GREAT EASTERN RAILWAY.

NORTH AND SOUTH WOOLWICH FERRY.

IN the late 'forties the Eastern Counties Railway—later becoming the Great Eastern Railway—started a ferry service between North and South Woolwich. They employed two paddle steamers. The fare for many years was one penny in each direction but the establishment of the ferry aroused considerable antagonism from the Watermen who possessed the ancient right of carrying passengers. This opposition was overcome by the railway company employing Watermen as masters and mates of the ferry steamers.

On the Essex side of the river the railway company built a large and spacious floating pier and a steamer was always ready to carry passengers across as soon as a train arrived from Fenchurch Street, Liverpool Street or Palace Gates as the case might be.

In connection with the up trains, the boats left the Kent side of the river ten minutes before the advertised time of departure of the trains from North Woolwich, and as the fares to Town were considerably cheaper than those charged by the South Eastern Railway, advantage of this route was taken by many people going regularly to the

P.S. " Middlesex " (1879)

Woolwich Ferry Boat " Gordon " (1923)

City. The main disadvantage of this route was the fog during the winter months.

For several years the railway ferry boats worked side by side with the London County Council free ferry but the competition was an unfair one and the bulk of the local traffic was transferred to the boats that worked at the expense of the rates, and as in the majority of cases, through tickets were issued from South Woolwich at the same fare as from the station on the north side, the railway ferry was run at a considerable loss, and it was not surprising that the G.E. eventually abandoned the service. The closing took place on 1st October, 1908.

The Great Eastern Railway also ran another Thames ferry service. This was between Brunswick Wharf, Blackwall and Greenwich, which commenced in July, 1866, and was discontinued on 1st November, 1902. Particulars of the ferry boats were as follows:—

ESSEX and KENT.

Built at Barking 1847. Length 78.5 ft. by 14.9 by 7.3. Gross tons 65. N.H.P. 32. Built of wood.

RAINBOW.

Built and engined by J. Stewart, London, 1864. Length 90 ft. by 18.5 by 9.9. Gross tons 112. Built of iron. Two cylinder lever engines, cylinders 30 in. by 42 in.

MIDDLESEX.

Built 1879 by the Thames Iron Works, Blackwall. Length 100.3 ft. by 18.7 by 7.7. Gross tons 103. Built of iron. Engines by T. A. Young & Sons, London. Two oscillating cylinders 25 in. by 30 in. N.H.P. 36. Converted to a hulk 1908.

WOOLWICH.

Built 1890 by the Thames Iron Works. Length 100 ft. by 26 by 8. Built of steel. Engines by Young & Sons. Two oscillating cylinders 25 in. by 30 in. N.H.P. 40. Went to Bowness after the closing of the ferry.

BRITISH RAILWAYS.

TILBURY AND GRAVESEND FERRY.

There had been facilities for crossing the river between Tilbury and Gravesend since Saxon times, but the first ferry boats were introduced about 1855. The service was taken over by the London, Tilbury and Southend Railway a few years later, then passed to the Midland Railway (1912), the L.M.S.R. (1923) and British Railways (1948) as each was in turn absorbed by a greater organisation.

The ferry provides conveyance for all types of vehicles, but prior to 1893 only passengers were conveyed.

The past and present ferry steamers are as follows:—

Name	Date	Builders	Mat.	Type
CATO	1849	Fawcett, Preston & Co., Liverpool	Iron	Pad.
EARL OF ESSEX				
EARL OF LEICESTER				
*SIR WALTER RALEIGH	1855	J. Henderson & Co., Renfrew	Iron	Pad.

* Originally named TILBURY. Cut up 1905.

The Thames Ferry Services

Name	Date	Builders	Mat.	Type
¶Thames	1868	Fawcett, Preston & Co., Liverpool	Iron	Pad.
†Tilbury	1883	J. & K. Smit, Rotterdam	Steel	Pad.
Carlotta	1893	A. W. Robertson & Co., London	Iron	T. Screw
Rose	1901	do.	Steel	T. Screw
Catherine	1903	do.	do.	do.
§Gertrude	1906	do.	do.	do.
Edith	1911	do.	do.	do.
Tessa	1924	Lytham Shipbuilding Co.	do.	do.
Mimie	1927	Ferguson Bros., Port Glasgow	do.	do.

Name	Dimensions	Gross Tons	N.H.P.	Cyls.	No.
Cato	115.2 × 18 × 7.7	128	56	30 in. × 36 in.	2*
Earl of Essex					
Earl of Leicester					
Sir Walter Raleigh	128.9 × 20 × 8.1	180	50	26.5 in. × 33 in.	2*
Thames	106 × 20.1 × 8.6	125	70	34 in. × 42 in.	2*
Tilbury	140 × 22.1 × 8.2	269	70	34 in. × 46 in.	2*
Carlotta	124.7 × 32.5 × t.7	261	50	12, 24 in. × 18 in.	4
Rose	124.7 × 24.7 × 7.8	259	100	12, 24 in. × 18 in.	4
Catherine	124.7 × 24.7 × 7.8	259	70	12, 24 in. × 18 in.	4
Gertrude	125 × 26.7 × 7.9	255	48	12, 24 in. × 18 in.	4
Edith	131.6 × 28.5 × 8.4	283	48	12, 24 in. × 18 in.	4
Tessa	135.1 × 30 × 10.5	371	46	11, 24 in. × 18 in.	4
Mimie	145.3 × 39.1 × 14.2	463	45	11, 22 in. × 18 in.	4

All machinery made by ship's builders excepting that for Tilbury which was made by Penn, Greenwich, and that for Sir Walter Raleigh by Henderson, Colbourn, Renfrew.

The Thames Steam Ferry Company.

The Thames Steam Ferry Co. operated from the Tunnel Wharf—now Middleton's Wharf—Wapping on the Essex side to a site adjacent to Church Stairs, Rother-

* Oscillating engines. ¶ Cut up 1913.
† Cut up 1922.
§ Purchased 1932 by New Medway Steam Packet Co. and renamed Rochester Queen.

hithe on the Surrey side of the river. The ferry had been organised for the purpose of supplying a much-felt want of a below-bridge communication between the north and south sides of the river.

The first pile of the ferry pier was driven with elaborate ceremonial on the 11th May, 1875. When the Lord Mayor of London in his robes of state accompanied by Members of the Corporation proceeded to All Hallows Pier, where they were received by Mr. Edmund Waller—Managing Director of the Company—on board of the PRINCESS ALICE which was decorated with flags.

On arriving at the temporary stage erected at Wapping which was gay with flags and streamers, the Lord Mayor was loudly cheered by the crowd.

The Rev. Canon McCaul, the Lord Mayor's Chaplain, having offered up a prayer, his Lordship turned the handle of the crab-winch or windlass and then declared the first pile of the pier of the Ferry Company to be " properly screwed." The first trial trip took place on 29th March, 1876, but the ferry was not opened to the public until November, 1877, the Lord Mayor once again formally attending. The service continued in operation until 1886.

The service was run by two paddle boats named the JESSIE MAY and the PEARL, the former built in 1876 and the latter in 1877. The boats ran either end first and were designed to convey horsed vehicles as well as pedestrians. Both boats were built by Edwards & Co., London, and engined by Maudslay, Sons & Field. The JESSIE MAY was of the following dimensions:—Length 79 ft. by 27 by 8.

Gross tons 149. N.H.P. 30. Two diagonal cylinders 23 in. by 27. The PEARL 82.9 ft. long by 27.2 by 7.2. Gross tons 120. N.H.P. 30. Two horizontal cylinders 23 in. by 27.

THE GREENWICH FERRY COMPANY.

The North and South Greenwich Ferry was opened in 1888. This small Company owned one boat, a double-ended twin screw built by Stewart & Lathom of Millwall, named the COUNTESS OF LATHOM. This steel-built ferry boat was in dimensions 120 ft. long by 40 by 10.1 and had a gross tonnage of 338. She had two sets of compound engines by Appleby Bros., London, having cylinders 19 in., 34 in. by 24 in. giving a N.H.P. of 94.

About 1891 or 1892 the Service was withdrawn, and the ferry-boat was sold in 1894 to a Portsmouth concern known as the Southsea and Hayling Steam Ferry Co. The boat was then renamed HAYLING. This service was given up in 1902 when the HAYLING was sold to the Nordischer Bergungs-Vreein at Hamburg, who converted her to a salvage vessel and renamed her UNTERELBE. The gross tonnage was increased to 626.

THE LONDON COUNTY COUNCIL FREE FERRY.

NORTH AND SOUTH WOOLWICH.

The Woolwich Free Ferry of the London County Council for the conveyance of passengers and vehicles was opened for public use on 23rd March, 1889. A regular nine minutes service is maintained between North and South Woolwich from 5-0 a.m. until midnight. The ferry boats employed are of ungainly appearance, and their action reminds one of a crab, only for a short distance in midstream do they progress as other vessels. Most of the 1,050 feet journey is accomplished by side-way kicks of their paddle wheels as they push off from or sidle up to the pontoons. The sidling is accomplished with such accuracy that the four hydraulically controlled fallbridges can be lowered immediately into place for the passengers and vehicles to embark and disembark.

The average number of vehicles carried is about 16,000 a week, which is rather less than 2,290 a day, including Sundays. The ferry boats can carry 100 tons of vehicles, on one ordinary occasion the author counted 5 delivery vans, 2 large 15-ton lorries, 3 horse-drawn coal carts, 2 petrol wagons, 8 private motor cars, several pedal cycles and 5 motor cycles.

The machinery of these boats consists of two entirely independent sets of simple condensing engines, each with two opposed cylinders working its paddle entirely independently. They have a speed of 8.5 knots when the river conditions permit them to utilise it.

The fleet at the present time consists of four vessels:—

Name	Date	Builders	G. Ton	Cylinders
SQUIRES	1922	J. Samuel White & Co., Cowes	609	(4) 33 in. × 16 in.
GORDON	1923	do.	609	do.
JOHN BENN	1930	do.	621	do.
WILL CROOKS	1930	do.	621	do.

The three original vessels as follow have been withdrawn:—

Name	Date	Builders	G. Tons	Cylinders
Duncan	1889	R. & W. Green, Millwall	493	(4) 33 in. × 16 in.
Gordon	1889	do.	493	do.
Hutton	1893	Simons & Co. Renfrew	554	do.

Gordon and Duncan - Length 164.6 ft. × 42.1 × 6.3
Hutton - - - Length 170 ft. × 42.1 × 6.3

The Gordon made the initial trip when Lord Rosebery declared the ferry open to the public; it was the first free ferry in England. Needless to say the ferry was a boon to the school boys at Woolwich on half holidays not excluding the author.

The Woolwich Arsenal Ferry.

Gallions Pier, N. Woolwich and the Arsenal.

For a period during the first World War years when work on munitions at Woolwich Arsenal was at high pressure and thousands of workers were employed in shifts, there was a ferry operating between Gallions Pier and a temporary pier at the Arsenal.

Two paddle steamers were employed one being the Red Cross which was chartered from the Port Sanitary Authority at Deptford and the other a Wallasey ferry steamer from the Mersey, the John Herron.

CHAPTER X.

MISCELLANEOUS STEAMERS.

EXPRESS.

During the summer season of 1884 there commenced running from Blackwall Pier to Gravesend, Southend, Sheerness and The Nore a new screw steamer named EXPRESS. Registered at Lloyds, unclassed, in the name of Loftus Perkins, but in 1886 in the joint names of Loftus Perkins and William Finlay, she was licensed to carry 677 passengers (river) and 345 (sea). Her length was 158 ft., breadth 20 ft. and the extreme depth 9 ft. 4 in. with a gross tonnage of 142. Rigged with a single mast forward, her one funnel was painted red with a black top, while the hull was painted black, below the water line red and lined out in yellow below the bulwarks. Accommodation consisted of an upper deck and a cabin deck with a fore saloon and a first class saloon aft of the engine room.

The hull, built of steel and fitted with six water-tight compartments, was constructed by the well-known Thames firm of shipbuilders, Samuda Bros. of Poplar in 1883 and the engines and boiler of high pressure design by A. M. Perkins & Son of Regency Square, Gray's Inn Road, London. Consequently, the EXPRESS was of especial interest for three reasons. First, she was one of the first, if not the first, screw excursion steamer on the Thames and was originally fitted with two screws, one in the stern and the other in the bow which was removed at a later date; Secondly, her engines were quadruple compound having a triangular connecting rod driving on to one crank thereby eliminating dead centres; and Thirdly, her water tube boiler, after having been passed by the Insurance Co. under an hydraulic test of 1,200 lbs. per square inch, worked at the then extraordinary high steam pressure of 400 lbs. per square inch.

The engines, placed amidship, had four cylinders, each steam jacketted, 7 inches, 12 inches, 14.75 inches and 25.5 inches respectively, the two former having a piston stroke of 9 inches and the two latter of 11 inches, respectively, running at 150 r.p.m. giving a speed of 10-12 knots. The valves were of the mushroom drop type operated by Stephenson's link motion. The piston rings were made of Perkins' Patent Metal which eliminated the use of lubricating oil in the cylinders, thus keeping the exhaust steam free from oil contamination before being condensed in the condenser for re-circulation and pumping back into the boiler from the hot well. The make up of distilled water was evaporated from fresh water in an high pressure still in order to maintain the level in the hot well. The usual waste steam pipe was not fitted alongside the funnel, the safety valves when blowing off passed the steam direct into the condenser.

The design of her engines proved unsuitable for the steel hull which had to be strengthened with wood, reducing her bunker capacity from 10 to 7 tons of fuel—coke and anthracite. Resulting from the employment of high pressure steam the fuel consumption was 1 lb. per indicated horsepower per hour as compared to 2½—3 lbs. per hour of similar craft employing low pressure steam.

Although the EXPRESS with her fine lines sailed during the summer seasons of 1884-88, first under the command of Captain R. G. Mathews and later under Captain W. Kirby, her performance must be regarded more in the nature of an experiment. Her machinery was undoubtedly fifty years before its time and demonstrated in an age not then appreciative of the fuel economies that could be obtained with steam at high pressure, but now indeed rendered necessary by the changed economic position of the present day. Yet the watertight compartments may have been a tragic reminder of the loss of the PRINCESS ALICE in 1878.

As a paying proposition for excursion service her speed was too slow and screw propulsion having compared unfavourably with paddles for quick and easy manoeuvring in a river then crowded with sailing and other small craft, it was decided in 1889 to withdraw her from service.

In 1890 William Taylerson—one of the proprietors of the EXPRESS—sold her to Andrew Ogle a shipbroker of Liverpool. In 1893-4 the British owners were deleted from the Register Book and the vessel was transferred to the Norwegian flag. The vessel was renamed SOLSKIN (Sunshine) in 1895-6. Owners: Chr. Michelsen & Co. Port of Registry: Bergen, where the strength of her hull no doubt gave her an added advantage in these waters.

In this connection it may be stated that with engines of novel design and a high pressure water-tube boiler containing very little distilled water when under steam, both the First and Second Engineers were specially trained by Messrs. A. M. Perkins & Son, for duty on the vessel. Consequently it is not altogether surprising to record that new propelling machinery, of which no particulars are available was fitted, made by Mjellem and Karlsens M.V. of Bergen for the new owners.

She then became the pleasure yacht of Chr. Michelsen and having been sold many times since his death, it is understood that she was wrecked during the World War, 1939-45, although her name disappears from the Register Book in 1896-7.

It is of interest to note that Loftus Perkins (1834-1891) was the grandson of Jacob Perkins (1766-1849) the patentee of the high pressure steam machine gun demonstrated before the Duke of Wellington and the Board of Ordnance in 1825, the " Uniflow " high pressure steam engine in 1827 and as founder of Perkins, Bacon & Co. made the steel dies and printed the first postage stamp in 1840.

A. M. Perkins & Son was founded by Angier March Perkins (1799-1881) a son of Jacob Perkins and father of Loftus Perkins. This Company is now incorporated in Baker Perkins Ltd., of Peterborough, manufacturers of the Perkins' Patent Steam Oven known and used the world over for baking bread.

William Finlay (1816-1889) was consulting and resident engineer at Gainsborough to Peto & Betts, railway contractors for the construction of the Manchester, Sheffield & Lincoln Railway in 1848, founder of the Dorking Grey-Stone Lime Co. Ltd. of Betchworth in 1865 and of the Gypsum Mines Ltd., Battle, in 1875. In both these works A. M. Perkins & Son's high pressure engines and boilers provided all the stationary and winding power then used. William Taylerson (1852-1924) was a nephew of William Finlay.

RACOON and CLAUD HAMILTON.

FOR many years there was to be seen lying off the Deptford Foreign Cattle Market one and sometimes two earlier cross-Channel paddle steamers: these were the two-funnelled RACOON and the CLAUD HAMILTON. The former a one time Burns line steamer plying between the Clyde and Belfast which was purchased in 1882 by the Corporation of London for transferring cattle at Gravesend from the large cattle boats that drew too much water to come up the Thames as far as the cattle market at Deptford. The RACOON was thus employed until 1907 when she was withdrawn and sold for breaking up. Her dimensions were as follows:—Length 252.3 ft. by 27.8 by 14.2. Gross registered tonnage 831. Engines, two oscillating cylinders 61 in. by 72 in. N.H.P. 300. Built 1868 by J. & G. Thompson, Glasgow.

The CLAUD HAMILTON was a former Great Eastern Railway boat and was the last paddle steamer to be employed on the Great Eastern Railway Continental Service. In 1897 she was purchased by the Corporation for the same purpose as RACOON and continued in service until 1914 when she was withdrawn and sold to Dutch shipbreakers. Her dimensions were as follows:—Length 251.6 ft. by 30.2 by 13.7. Gross registered tonnage

S.S. " Express " (1884)

P.S. " Racoon " (1868)

962. Engines, two compound oscillating cylinders 54 in., 95 in. by 63 in. N.H.P. 350. The CLAUD HAMILTON was built in 1875 by J. Elder & Co., Glasgow.

ARRAN.

The ARRAN was built in 1867 as the DUNOON CASTLE for the Dunoon and Rothesay Carriers, but she was not in service for long and for many years was laid up. In 1883 she was purchased by Hill & Co. for the Fairlie & Millport service. Two years later she was acquired by Campbell & Gillies who ran her in connection with the Wemyss Bay Co. Eventually she was taken off the Clyde and went round to the Mersey for a period and then to the Thames.

The vessel was the last of the Clyde steamboats to be fitted with a steeple engine and also the last Clyde steamer to have the funnel placed aft of the paddle-boxes. Later she came out with two funnels both aft of the engines—an unusual arrangement—later the second funnel was removed and at the same time the name was changed to ARRAN.

Built and engined by T. Wingate & Co., Glasgow. Dimensions: Length 191.7 by 18.2 by 7.5. Built of iron. Gross tons 171. One cylinder 50 in. by 39 in., later re-engined with a cylinder 50 in. by 45 in. N.H.P. 85.

The ARRAN worked only one season (1891) on the Thames under the ownership of the Thames & Medway Saloon Steamboat Co. (Mr. Pursell), under Captain Stephen Jones. She sailed from Fresh Wharf on Mondays, Tuesdays, Wednesdays and Saturdays to Southend,

Sheerness, Gravesend and Rosherville Gardens. On Thursdays and Sundays she made trips round the Nore Lightship. The ARRAN finally finished up as a tug on the Shannon.

QUEEN ELIZABETH.

The QUEEN ELIZABETH was a steel built paddler constructed in 1895 by Edwards & Co., Millwall, for a Mr. Edgar Shand. She was of 141 gross register tons and the dimensions were: Length 159.3 ft. by 15.5 by 6.5. She had a two-cylinder compound engine built by Wilson & Co., London. The QUEEN ELIZABETH ran between London Bridge, Kew, Richmond and Hampton Court. Her full compliment was no less than 709 passengers. In March, 1896, she was chartered to take the guests of the Thames Iron Works to witness the launch of the Japanese battleship FUJI. About 1897 she was transferred to the Queens of the River S.S. Co. of London. In 1913 she was acquired by Mr. J. T. Mears, Richmond, for his river Thames trips. At the close of the first World War QUEEN ELIZABETH was scrapped.

VICTORIA.

The VICTORIA was a handsome two-funnelled boat having fore and aft main deck saloons. She was built in 1886 by Blackwood & Gordon, Port Glasgow, for the Wemyss Bay Steamship Co.

In 1890, she changed ownership being purchased by Mr. A. Campbell and in the following year was taken over by the Scottish Excursion Steamers Co. who ran her until she came round to the Thames in 1895 having been purchased by the London & East Coast Express Steamship Co. who ran her from London to the East Coast resorts until the close of 1900.

VICTORIA was of the following dimensions: Length 222.4 ft. by 23.1 by 8. Gross tons 341. Two cylinder diagonal engines, the cylinders 40 in. by 66. N.H.P. 200.

CONQUEROR.

Whilst the LA MARGUERITE was leading the way in Continental day excursions from London, another and quite small vessel was pioneering in a different direction. This was the CONQUEROR of 1897, which must not be confused with a similar boat named CONQUEROR of 1884, unlike most excursion steamers in recent times she was privately or semi-privately owned by Messrs. Dick & Page (Elliott Steam Tug Co., London). During the summer months the CONQUEROR was rigged out as an excursion boat and in the winter months was detailed for towage duties.

This little steamer opened up an important chapter in the history of Thames excursion facilities as she was the pioneer of the seasonal excursion service from London to Dover. Other and larger steamers were to follow in her wake, but it is to the CONQUEROR that the distinction belongs of being the first excursion boat from London to disembark passengers at Dover. The exact date is uncertain—some time between 1900 and 1904. For a time the CONQUEROR was stationed at Margate as a base for running coastal trips from that resort. Later she ran from

Southend and Sheerness. She was remarkable inasmuch as she was two funnelled for such a comparatively small vessel. Dimensions: Length 131 ft. by 21.7 by 10. Built of iron, 1897, by J. T. Eltringham & Co., South Shields. Gross tons 224. Two cylinder side lever engines, having 31.25 by 58 in. cylinders. N.H.P. 150. Running until 1925.

ENGADINE.

Commencing on the 1st June, 1932, Messrs. S. Instone & Co., London, opened a new daily service—Sundays included—from the Tower Pier to and around the Nore lightship. This service was made by the ENGADINE, a Southern Railway cross-Channel turbine steamer, which had been chartered for the season by the promoters of the trips.

The ENGADINE was scheduled to leave London at 10.45 a.m. and returning back about 5 p.m., these times enabling passengers from outlying districts to board the steamer in no undue hurry and return at a reasonable hour, but owing to insufficient patronage the ENGADINE did not run a second season. The ENGADINE was a triple screw turbine vessel of 1,676 gross register tons, 316 ft. in length by 41.1 by 15.8, built in 1911 by Denny Bros., Dumbarton. She was a vessel not without history being the last boat bringing passengers from Belgium before war broke out arriving at Dover at 8 a.m. on the 3rd August, 1914, with 18 passengers. In December, 1914, the ENGADINE was taken over by the Admiralty and during the War acted as a seaplane carrier and had the distinction of being the first ship to sight the German fleet at the Battle of Jutland,

P.S. " Showboat " (" Alexandra ") at Shoreham in 1933

Face page 164

T.S.S. "Lady Enchantress" in 1947

and she was the first seaplane carrier of any navy to take part in a fleet action.

One of the ENGADINE's outstanding achievements was the rescue of the crew of H.M.S. WARRIOR sunk at Jutland on 1st June, 1916. In 1934 the ENGADINE was sold to the Compania Maritima, of Manilla, when she was renamed CORREGIDOR and worked between two of the Philippine islands, a distance of 350 miles. Her career ended in the second World War when she struck a mine in Manilla Bay.

NEW DAGENHAM.

This handsome craft was built at Alblasserdam, Holland, in 1933, and was in commission the same year running trips from Westminster, principally to the Ford Motor Works at Dagenham, until 1937 when she was sold to Turkish owners. Driven by two Diesel engines of 200 b.h.p. each, she was 134 ft. long and had a breadth of 22.1 ft., depth 9 ft 6 in., and a gross tonnage of 255. Her owners were the Odelia Cruising Motor Ship Co., Ltd., a subsidiary of R. G. Odell, Ltd., the well-known firm of lightermen and tug owners.

SHOWBOAT (ALEXANDRA).

In the early summer of 1932 another innovation was introduced on the Thames. The novelty was the ex South Coast paddle steamer ALEXANDRA which had been converted to a luxury floating restaurant, dance hall and cabaret, and was advertised as SHOWBOAT. Every evening SHOWBOAT left Westminster Bridge Pier to make a dinnertime cruise down the river. The cruise on the return trip ended at the starting point about 11 p.m. and SHOWBOAT then proceeded up the river to Richmond for a supper cruise.

Although the boat was licensed to carry 600 persons, for the sake of comfort, the number was limited to 200. The experiment, however, was not a success with the London public and she only worked one season on the Thames. In the spring of 1933 SHOWBOAT was towed round to Margate and anchored off the town, brilliantly lighted at night and dancing with a cabaret show was provided. Towards the end of July her venue was again changed and her attractions were offered to the residents of Bungalow Town, Shoreham, the boat being moored in the river Adur where she was easily accessible to holiday-makers at Brighton. With her engines disconnected and her paddle wheels slowly revolving with the tide, the old veteran remained at Shoreham until 26th October, 1934, when she was taken in tow for Grays to be broken up.

The SHOWBOAT was officially named ALEXANDRA, Weymouth being her port of registry. Built in 1879 by Scott & Co., Greenock, to the order of the Portsmouth & Ryde United Steampacket Co. to run between Portsmouth and Ryde. In the year following the Company was taken over jointly by the L.B. & S.C. and the L. & S.W. Rys. Later the joint owners sold the ALEXANDRA to the Bembridge & Seaview S.S. Co. In 1915, she was purchased by Cosens & Co., Weymouth, but in September of that year was taken over by the Admiralty, renamed ALEXIS and continued in the Government service until February, 1920, when she was sold and subsequently purchased by a London syndicate to become the SHOWBOAT. The ALEXANDRA was a vessel of 235 tons gross, 171 ft. in length by 20.2 by 8.5 and had compound diagonal engines of 153 N.H.P.

LADY ENCHANTRESS.

Another excursion steamer venture was started on 4th August, 1947, with the twin screw steamer LADY ENCHANTRESS, a luxury ship run by the Three Star Shipping Co. of Ramsgate. The service operated from Gravesend, calling at Southend and Margate on Tuesdays, Thursdays, Saturdays and Sundays and then from Margate for a cruise off the French Coast within a short distance of the famous Dunkirk Beaches, whilst on Mondays and Wednesdays she worked a Kent Coastal trip and off the Goodwin Lightships.

The LADY ENCHANTRESS is a ship with a history: she was laid down on 9th March, 1934, as the BITTERN, a naval escort vessel, at the yard of John Brown & Co., Clydebank. While under construction the original intention was abandoned and she was launched on the 21st December, 1935, as an Admiralty yacht carrying the name ENCHANTRESS. The vessel cost about £150,000 when new, minus armament. During the War the ENCHANTRESS travelled many thousand miles, the last time she was in the news during the War in the Mediterranean was when she rammed a U boat, folding back 15 ft. of her bows and had to be towed to Oran to have new bows fitted.

The work of conversion was undertaken by Messrs. J. I.

Thornycroft & Co., Southampton. The LADY ENCHANTRESS, which is painted white, carries two masts, has two main and one sun deck. Her main dimensions are as follows: Length 269.4 ft. by 37 by 15. She has a cruiser stern and is of 1,474 gross tons. The propelling machinery consists of four geared turbines. She only worked one season on the Thames and since has remained moored out of commission at Gravesend.

EDWARD WILLIAM and MYLETA.

These two paddle boats, owned by the South Eastern Railway, were employed on a service between Port Victoria and Sheerness between 1891 and 1897. They were both the product of Samuda's Yard, and were of 195 gross tonnage; length 150 ft., beam 18.6 ft. and depth 8.2 ft.

METROPOLITAN ASYLUM DISTRICT BOATS.

Some few years back one could see down the London river several paddle steamers belonging to the Managers of the Metropolitan Asylum District tied up on the South side of the river off Rotherhithe. These vessels were old water ambulances out of commission, having been originally employed in carrying patients suffering from infectious diseases from the clearing house at Rotherhithe to the isolation hospital at Dartford. They had black funnels with a brown band round.

In 1932, the craft was taken over by the London County Council and eventually sold for £400 the lot. Today patients are conveyed by road ambulances. The boats were as follows:—

ALBERT VICTOR.

Built 1879, by Messrs. Westwood, London. Length 129 ft. by 17.5 by 6.7. Gross tons 106. Engines by the London Steamboat Co. Two oscillating cylinders 23 in. by 30. N.H.P. 36.

MARGUERITE.

Built 1879, at Cubitt Town. Length 68.1 ft. by 10.1 by 4.5. Gross tons 23. N.H.P. 10.

MALTESE CROSS.

Built 1894, by Edwards & Symes, London. Length 132 ft. by 16.6 by 6.7. Gross tons 134. Engines by Pratt, Huddersfield. Two oscillating cylinders 23 in. by 30. N.H.P. 35.

GENEVA CROSS.

Built 1894, by J. Stewart, London. Length 137 ft. by 22.1 by 6.7. Gross tons 260. Compound engines by builders, cylinders 23 in., 46 in. by 42. N.H.P. 114.

RED CROSS.

Built 1901, by Mordey & Carney, Southampton. Length 136 ft. by 18.6 by 8.4. Gross tons 162. Compound engines by builders, cylinders 24 in., 40 in. by 36.2. N.H.P. 72. The boat was loaned to the Government from 1917 to 1919.

EXMOUTH II.

The Managers of the M.A.D. also had a three-masted schooner—classed as a training ship—fitted with an auxiliary engine. This vessel was named EXMOUTH II, built in 1913, by Rennie, Forrest & Co., Wivenhoe. Dimensions: Length 146.3 ft. by 27.1 by 13.3. Gross tons 350. It was fitted with a two cylinder compound engine made by A. G. Mumford & Co., Colchester. N.H.P. 24.

The above-mentioned isolation hospital at Dartford consisted of three vessels lying in that stretch of the Thames known as Long Reach; they were the old line-of-battle ship ATLAS, the frigate ENDYMION, and the unique twin paddle steamer CASTALIA which was fully described in the author's "Cross-Channel and Coastal Paddle Steamers." The CASTALIA was towed to Dordrecht in Holland for breaking up in 1907.

* * * *

A number of other boats have worked for short periods on the Thames. One was the BONNIE DOON, "a Clyde built steamboat lighted by electricity." She ran for a few weeks from Fresh Wharf to Clacton, but her owner was soon in financial difficulties. In a police court case Richard Ford, ship charterer, answered several summonses. Samuel Middleton stated that he had been engaged by defendant to meet the BONNIE DOON, swing her, and moor her. He had also to do other work connected with the ship and his wage amounted to thirty-six shillings. Captain Mason stated that he was engaged by the defendant from July 29th to October 15th at a wage of £7 a week. Defendant owed him three weeks wages. Joseph Haw said he was engaged as chief steward at £2 a week and when the ship stopped he was to receive £10 gratuity. The ship had stopped running because the merchants refused to supply her with coal as the coal bill was not paid. The magistrate made orders for the various amounts claimed to be paid.

An old handbill, dated August 23rd, 1872, advertises a boat called PALMERSTON as running a special trip from Old Swan Pier to Gravesend to see the American fleet; she was possibly one of Mears' boats. Another handbill, dated 1889, advertises the HERALD, "the fastest boat on the Thames," running from Blackwall to Margate and Boulogne.

CHAPTER XI.

THE WILLIAM WATKINS PADDLE TUGS.

ALTHOUGH tugs are not strictly relevant to the subject of this book, the reason for including a chapter on the oldest English firm of tug owners is that several of these have been used from time to time on seasonal excursion and other passenger work.

In years gone by, the tug was a very important vessel, but since the disappearance of the sailing ship, its former chief activity is now very seldom called for, and today the work of tugs is confined to towing barges and assisting big ships in and out of dock.

The first steam tug that came to the London river was the LADY DUNDAS in 1832; she came from the Tyne and was followed soon after by the WEAR which came from Sunderland. The latter was a failure on the Thames, not being sufficiently powerful to work against the tide, much less to take a ship in tow.

In 1833 the firm of W. Watkins was founded and this company is the only one of the pioneer English tug-owners to survive, although in Scotland the Clyde Shipping Co., still one of the most important tug owners in Britain, was started in 1819. On the Thames the Watkins tugs are conspicuous with their black and green hulls, white paddle boxes and black funnels with a deep red band painted well down from the top.

Monarch, 1833.

The first tug owned by Watkins was the Monarch, needless to say a paddler; with two exceptions all the firm's tugs were paddlers until 1880, when the screw was finally adopted. The Monarch hailed from the North. Built of wood and nearly 65 feet in length, 14 ft. beam and a draught of 7 ft. 5 in. She was clincher built, having a cut-away bow and counter stern. Tonnage just over 26. The engine registered 20 N.H.P. being built by Wait of N. Shields. In 1845, the original flue boiler was replaced by a tubular boiler which greatly reduced the fuel consumption as the original boiler was stated to have been a regular " fuel eater."

In 1856, in spite of her age, new patent paddle wheels were fitted and the engine, boiler and hull were overhauled, at a cost of £1,000, by Wigram's at Blackwall.

In 1861 the Monarch was again given a new boiler, and in 1876, she was sold for scrapping, the sale realising £40.

Fiddler, 1840.

In 1840 came the Fiddler, but no details of this tug have survived, beyond the fact she could take on jobs that were beyond the capabilities of the Monarch.

Lord Warden.

The Lord Warden followed the Fiddler, but little is known about her, except that her engine was put into another tug when her hull was worn out.

PUNCH (1), 1846.

In 1846, the PUNCH was acquired, afterwards she went out to Constantinople, being one of the first tugs in Turkish waters.

PAUL PRY, 1847.

The PAUL PRY was built at Millwall in 1847, clincher built and 80 ft. in length with a beam of 15 ft. 5 in. and a gross tonnage of 83. In 1858, she was overhauled and fitted with a new boiler and patent paddles which gave her another lease of life, as the hull was perfectly sound, although she had been laid up since the close of the Crimean War as obsolete. In 1865, she was in again for general overhaul and in the year following was sold to the Commercial Towing Co., but after that did not survive on the river long.

JOHN BULL, 1849.

In 1849, JOHN BULL was added to the fleet, she was carvel built by Wigram's, Blackwall. Length 96 ft. having an 18 ft. 2 in. beam and a gross tonnage of 114. In 1853, Mr. Watkins Junr. who was setting up in business as a tug owner at Birkenhead on the Mersey took over the JOHN BULL. In 1859, the JOHN BULL returned to the Thames and then had new side-lever engines and boilers installed.

UNCLE SAM, 1849.

The UNCLE SAM came in 1849, built by C. J. Mare & Co., West Ham, 101 ft. 7 in. in length by 18 ft. 3 in.

beam. This veteran survived until 1900, when she was sold to Dutch shipbreakers.

SAUCY JACK/JOHN LEE, 1850.

This tug was built in 1844 by A. Woodhouse of South Shields. She was clincher built 90.2 ft. long by 17 ft. beam with a registered tonnage of 28. When she came round to the Thames she was not very successful and in consequence was sold to the Southampton Dock Co. for handling big steamers. In December, 1850, she was purchased by the Watkins firm and re-named SAUCY JACK, but after a time she reverted to the original name of JOHN LEE.

BRITANNIA (I), 1852.

In 1852, the BRITANNIA was built by Money, Wigram & Co. at their Northam shipyard. This tug was a wooden paddler 100.5 ft. in length having a beam 16.7 ft. and carried a figure-head and bowsprit. It may be said there were two classes of Thames tugs, those mainly confined to the river and the sea-going fraternity. The sea-going tugs frequently went far down into the Channel and across to Continental ports; early in 1855 the BRITANNIA was put on this work and one of her first long tows was to take the 735 ton COLUMBIA from Deptford to Calais Roads. The BRITANNIA proved to be one of the best tugs on the Thames. Late in 1855 she was sold to Constantinople.

VICTORIA, 1853.

The VICTORIA was built in 1853, and was by far the largest and most powerful tug on the river for some years.

She was built of teak and oak and had a gross tonnage of 152. She had a clipper bow and one of her improvements was a steam windlass which saved a great deal of time in hauling in the anchor.

The VICTORIA was one of the few boats at the time to have two funnels athwart but soon after the practice of having two boilers side by side became general. In 1859, she was given new boilers and patent paddle wheels. However, the new boilers did not give satisfactory results and in 1864 another new set of boilers were put in and at the same time a general overhaul was given. A fourth set of boilers was put in in 1875. In January, 1906, this fine old tug was taken to Millwall for scrapping.

PUNCH (2), 1854.

In 1854, Money, Wigram of Blackwall launched the PUNCH—the second of her name. She was a fine little iron paddler of 115 tons gross drawing 6.5 ft. of water. At her bows she carried a fine carved wooden figure-head of Punch which she lost in a heavy sea in a gale in the Channel. However it was salvaged and afterwards graced the " Blue Anchor " public house in High Street, Newhaven until that hostelry was pulled down.

The PUNCH was the first mercantile vessel to pass under Tower Bridge after the opening in June, 1894. In November, 1895, this famous tug was sold to North Shields owners for £1,600 in spite of her age, and remained hard at work until November, 1911, when she was in collision off Shields, but although she was salved she never worked again and was scrapped.

P.S. " India " in 1890

P.S. " Cynthia " (1892)

Face page 176

DON, 1855.

Early in 1855, this little wooden paddler was purchased for river use only. She was clincher-built of 45.9 tons and 72.2 ft. in length with a 14 ft. 9 in. beam; a single cylinder engine provided the motive power. The DON was built as far back as 1841. She was broken up in 1865.

TOBY, 1855.

The composite built TOBY hailed from Greenwich in 1855. Length 80.3 ft. by 17 ft., and of 73 tons. In 1862, she was sold to Jersey, in 1879 she went over to France, and in 1882 she returned to England to go to Hull, remaining on the East coast until she was cut up in 1904.

DEFIANCE, 1856.

The DEFIANCE was an iron paddler built at South Shields. She commenced work on the Thames in 1856. In 1869, she was modernised and re-boilered, and in 1871 sold to the East and West India Dock Company.

NAPOLEON, 1857.

In 1857, the carvel-built paddler NAPOLEON was purchased, having been built at Southampton. This tug was specially designed for taking long tows and had all the latest refinements of her day. She had two funnels abreast and twin engines thus introducing the double engine principle on the Thames. Unfortunately her twin engines proved to be too powerful for her wooden hull resulting in continuous expense. The gross tonnage was

157. In July, 1870, the NAPOLEON was re-boilered and practically re-engined at Wigram's. In 1881 she foundered at sea.

ANTAGONIST, 1857.

The ANTAGONIST was an iron-built paddler by Lungley's of Deptford, built in 1857; a vessel 91.6 ft. in length. About 1863, the ANTAGONIST was chartered for a time by the Isle of Wight Steam Packet Co. to ply between Littlehampton and Ventnor. In August, 1867, she was sold to Dutch owners at Amsterdam.

VICTOR, 1859.

In 1859, the clipper stemmed tug VICTOR was bought, having been built at Deptford in 1857. To enable the VICTOR to undertake long distance tows, she was fitted in 1866 with a surface condenser, new boilers and the engines given a thorough overhaul. On one occasion, she towed a dredger from the Thames to Cadiz and on another occasion took a dredger to Ferrol from the Victoria Docks. In 1880, she was again re-boilered and considerable repairs to the hull and engines carried out giving her another lease of life; she lasted until 1894.

TIMES, 1861.

The TIMES, a wooden clincher-built paddler was purchased in 1861, being built at South Shields in 1857. She was a tug of 78 gross tons. In May, 1874, the TIMES was run down in the Thames, was salved and almost entirely rebuilt. She continued in service until broken up in 1887.

ANNETTE, 1862.

The ANNETTE, an iron paddler built in 1862 by James Ash of London. She was fitted with Stewart engines, having 29 in. diameter cylinders. The boilers were in advance of their time being fitted with superheaters and surface condensers instead of the old jet type. In the 1860's the ANNETTE was chartered by Messrs. Brett to run in a competitive service between Ryde and Stokes Bay. In 1869 the ANNETTE was sold and went to the Danube.

BRITANNIA (2), 1862.

The BRITANNIA, which had been built in 1855, was purchased by Watkins in 1862. A useful little paddler of 97 tons gross. She was sold in 1875 to the ship-breakers, the boilers being put into the EXPRESS.

EXPRESS, 1864.

In 1864, the EXPRESS was bought, an iron paddler of 94 gross tons, she was built in 1856 on the Tyne. Her dimensions were 88.9 ft. long with an 18.6 ft. beam. In 1876, the EXPRESS underwent a large refit and received the boilers taken out of the BRITANNIA (2) which had been repaired and classed as equal to new. In 1880, she was disposed of to owners at Milford where she worked until she was broken up in 1898.

ANGLIA, 1866.

The year 1866 saw the introduction of the very famous tug ANGLIA, the pioneer of all first-class long-distance tugs.

The ANGLIA was in many ways a very noteworthy vessel, firstly she had two funnels abreast abaft the paddle boxes and one forward of them and as such was known on the river as " Three Fingered Jack " and was always conspicuous at sea. The ANGLIA had three boilers, a funnel to each boiler; the original intention was to have four boilers and four funnels, an arrangement only equalled by the celebrated BEN-MY-CREE the large paddle steamer owned by the Isle of Man Steam Packet Co. at one time. The ANGLIA carried the largest bunkers ever put in a tug and she needed them as she was very extravagant on coal. Her boilers were tubular and the side lever engines developed 700 I. H. P. thus making her the most powerful tug on the London river and probably in the World. The ANGLIA was a fine sea-boat but she wanted careful handling in steering owing to her full lines aft. In 1872, she went to the Naval Review at Spithead held in honour of the visit of the Shah of Persia, conveying a party of guests. In 1880, she was thoroughly modernised and tackled some very big jobs, one being a tow to Malaga.

In 1894, the ANGLIA was sold as tugs of her date were obsolete, but it was not the end of her career as Messrs. Palmer's, the ship-builders, bought her. The engines, boilers and paddle wheels were scrapped and she was converted into a towing lighter to carry iron from the Tees to the Tyne.

When the 1914 war broke out she was employed working to and from the North East coast and France, generally in tow of a French warship, carrying coal for munition works and the railways. An event which made the ANGLIA conspicuous took place in 1875 and was referred to by every paper interested in shipping matters when she undertook the most difficult tow of the broken down Union Steamship Co. liner SYRIA from St. Helena to Southampton. Later she towed a dredger from Cardiff to St. Petersburg.

ALBION (I) *ex* FRIEND OF ALL NATIONS, 1868.

The paddle tug FRIEND OF ALL NATIONS was purchased in 1868 and almost at once renamed ALBION. This tug was built at Newcastle in 1851 and was of 168 gross tons. Her usual work was to steam down Channel to pick up sailing craft and go as far west as Land's End. Her time with the Watkins firm was short as she was underpowered for their requirements. She was sold in 1869 and went to Trieste.

CAMBRIA, 1870.

The CAMBRIA was built by the Thames Iron Works in 1870 and was a big improvement on the ANGLIA. She was fitted with two cylinder side-lever engines by Stewart and was a vessel of 209 gross tons; she had two funnels abreast, the mizen mast being placed close to the rear of the funnels but the foremast was fixed well forward. The CAMBRIA was distinguishable from the other paddle tugs as she had side houses on the front only of the paddle boxes. In 1876, the CAMBRIA was chartered by the Duke of Bedford to visit the yachting regattas round the coast and for this purpose she was painted white and the funnels yellow. With her clipper bows she was a handsome craft, and in view of long sea-going trips was built with large cabins enabling the crew to sling hammocks. In 1887, the CAMBRIA had the novel experience of being placed under arrest at Cherbourg for violating the French law. The CAMBRIA had towed a vessel from Dunkirk, the law being that all coastal trade must be undertaken by French vessels. The CAMBRIA, excepting the IONA, was the last remaining paddle tug on the London river.

ALBION (2), 1870.

The second ALBION was purchased in June, 1870, a small vessel of 109 tons, fitted with Stewart engines which gave a speed of 12 knots.

One of the ALBION's first duties was to cruise between Falmouth and the coast of Ireland to give warning to French ships of the outbreak of the Franco-German war. This duty was also performed by CAMBRIA.

Although a handy vessel the ALBION was not up to requirements as owing to her light draught she did not have a proper grip on the water when engaged on a heavy tow, so in January, 1872, she was disposed of to Belgian owners at Antwerp.

ROBERT BRUCE, 1872.

To replace the ALBION, the ROBERT BRUCE was purchased in 1872, a tug which had been built in 1865 for the Caledonian Steam Towing Co. at Poplar. The ROBERT BRUCE was a powerful iron-built vessel of 192 tons, a conspicuous feature was her very large paddle boxes. In 1892, she was laid aside and sold to the shipbreakers; she was not, however, broken up, but the boilers and engines were removed and the hull sold to the Port Sanitary Authority, London, who moored her off Gravesend as a medical hulk and named her HYGEIA. As such she remained until she was broken up in 1910.

ATLAS (I), 1873.

In 1873, a small wooden carvel-built paddle tug of 100 gross tons was bought from Bristol. She was getting on in years, as 1854 was her date, but in spite of that she was a handy acquisition. In 1879, the ATLAS had a thorough overhaul and was given two second-hand tubular boilers and continued in service until broken up in 1887.

HIBERNIA, 1874.

The HIBERNIA an iron vessel was added to the Watkins fleet in 1874, and was very similar to the CAMBRIA, she was fitted with side-lever engines and had a gross tonnage of 238. In 1883, the HIBERNIA was re-boilered and in the following year was sold to Havre and after many changes of ownership she returned to her original owners in 1911.

SCOTIA, 1874.

The SCOTIA was a similar tug to the HIBERNIA and with the CAMBRIA were regarded as Watkins' best tugs.

TITAN *ex* ATLAS, 1874.

Following the SCOTIA, in the same year, was purchased the Dundee-built tug ATLAS for river work; she was immediately renamed TITAN, as there was already a tug named ATLAS belonging to the General Steam Navigation Company. The TITAN was built in 1865. In 1888, the TITAN was sold to tug owners at North Shields where she spent many years, being specially suitable for the narrow river.

RENOWN, 1874.

In the same year, 1874, the RENOWN was purchased. She had been built at Deptford in 1863 and had Stewart-built engines and was an iron clipper bow paddler of 165 tons gross. The RENOWN was a light draught vessel. Prior to Watkins taking her over she had already seen service on the London river and prior to that was engaged at the Port of Fleetwood. In 1903, the RENOWN was sold to ship-breakers in Holland.

PILOT, 1875.

In the autumn of 1875, this wooden clincher-built paddler of 71 tons gross with engines of 30 n.h.p. was purchased from South Shields. This tug had been built as far back as 1858, and in 1864 had been fitted with new engines and boilers. A considerable amount of money was spent on her when purchased but she was a source of trouble nevertheless as the engines were too powerful for the hull and she was soon disposed of.

INDIA, 1876.

In 1876, Watkins placed an order with Westwood & Baillie for a sea-going tug, to be named the INDIA. She had a gross tonnage of 218 and was fitted with Stewart's side-lever engines. Unfortunately, lack of power was her main drawback. She was a fine sea boat and during the summer months of 1890-3 was employed on sea excursions running from Margate which proved a profitable undertaking. In 1894, she was sold to the Tees, and ten years later was turned into a hulk.

FOX, 1877.

The Fox was acquired in October, 1877, and disposed of in the following February.

MALTA, 1886.

In the year 1886, Watkins purchased three of the Ben tugs, the first being the small tug BEN ACHIE, which was renamed MALTA. This vessel did not remain long on the Thames. Whilst undergoing repairs at Newcastle in 1888 due to a broken paddle shaft she was sold.

BURMAH and IONA, 1866.

The other two Ben tugs, which were sisters, were the BEN LOMOND and the BEN NEVIS immediately renamed BURMAH and IONA respectively. When purchased they were in a poor condition and it was not until 1887 after they had been thoroughly overhauled that they were put in service. In 1899, the BURMAH was sold. The IONA had the distinction of being the last paddle tug to work on the Thames.

CYNTHIA, 1892.

In 1892, Messrs. Watkins influenced by the success of the INDIA in making sea trips from Margate, decided to have a steamer built especially for excursion work. So the CYNTHIA was built at South Shields by J. T. Eltringham & Co. and was an exceptionally handsome little vessel.

After running for some seasons from Margate, competition became too keen and she was withdrawn and for a time went North but came back later to the Thames for a while. For three seasons the CYNTHIA was running from Hastings and in 1930 she went to Ireland. In 1934 during a severe gale, whilst she was in Kingstown, she broke away from her moorings and was driven against the harbour wall becoming a total loss.

The advent of the screw tug in 1880 brings this chapter to a close.

The information in this chapter is largely taken from a privately issued souvenir, " A Hundred Years of Towage," written for the Watkins firm by Mr. F. C. Bowen.

INDEX

Index

Index

INDEX

Index